Biological Evaluation of Materials

The interaction of materials with their environment

Biological Evaluation of Materials

The interaction of materials with their environment

Bogdan Walkowiak
Department of Biophysics, Lodz University of Technology, Stefanowskiego st. 1/15, 90-537 Lodz, Poland, and Bionanopark Laboratories, Bionanopark Ltd, Dubois st. 114/116, 93-465 Lodz, Poland

Magdalena Walkowiak-Przybyło
Aflofarm Farmacja Polska Sp. z o.o. Partyzancka 133/15, 95-200 Pabianice, Poland and Institute of Materials Science and Engineering, Lodz University of Technology, Stefanowskiego st. 1/15, 90-537 Lodz, Poland

Piotr Komorowski
Bionanopark Laboratories, Bionanopark Ltd, Dubois st. 114/116, 93-465 Lodz, Poland and Department of Biophysics Lodz University of Technology Stefanowskiego st. 1/15 90-537 Lodz, Poland

IOP Publishing, Bristol, UK

ISBN 978-0-7503-2656-8 (ebook)
ISBN 978-0-7503-2654-4 (print)
ISBN 978-0-7503-2657-5 (myPrint)
ISBN 978-0-7503-2655-1 (mobi)

DOI 10.1088/978-0-7503-2656-8

Version: 20220901

IOP ebooks

British Library Cataloguing-in-Publication Data: A catalogue record for this book is available from the British Library.

Published by IOP Publishing, wholly owned by The Institute of Physics, London

IOP Publishing, Temple Circus, Temple Way, Bristol, BS1 6HG, UK

US Office: IOP Publishing, Inc., 190 North Independence Mall West, Suite 601, Philadelphia, PA 19106, USA

Contents

Preface

At first glance, materials science and the life sciences are so far removed from each other that it is difficult to find any connection between them, but such a relationship exists and is extremely important for further development. New materials with intentional properties, belonging to the group of biomaterials, allow for applications and medical technologies which have not been achieved until recently. However, with the current state of knowledge and technology, the dreams of creating one 'super biomaterial' suitable for all biomedical applications should be shelved. Even nature has not coped with such a challenge and for the performance of various functions it has developed different tissues with different mechanical properties and different profiles of biological activity, although the formation and functioning of all these tissues is based on one polymer material, which is the genetic material contained in the cell. This material is identical in every cell of a given organism, regardless of the function of the cell in the tissue, and the tissue in the body. In this sense, deoxyribonucleic acid (DNA) can be described as a 'super biomaterial'. Of course, in the future it may turn out that materials science is able to produce materials with unlimited ability to adapt to environmental conditions and thus able to replace different types of tissues in terms of both strength and function. To achieve this aim, close co-operation between the representatives of both materials science and the life sciences is necessary. Traditionally, the representatives of materials science usually use physicochemical methods of producing and researching new materials, including nanomaterials, usually not feeling even the smallest need to use biological methods in their research. The only exception is biomaterials engineering, where the need for biological assessment of biomaterials is obvious. It seems, however, that the era of biology in materials science has already arrived and the joint use of physicochemical and biological methods is key to progress and solving the most complicated problems in materials science. In this way, at the beginning of the twenty-first century, the field of materiomics, combining materials science with the life sciences, has appeared and emphasized the need for the biological assessment of materials.

Author biographies

Bogdan Walkowiak

Bogdan Walkowiak, a graduate of nuclear physics at the University of Lodz, Poland (1978), obtained his PhD in natural sciences from the Medical Academy in Lodz, Poland (1985). In 1989–90 he was a postdoctoral fellow at the State University of New York, Brooklyn, NY, USA. He was then employed at the Medical University of Lodz, Poland, at the same time receiving training at the companies Perkin Elmer, Pharmacia Biotech, Amersham Biosciences, BiaCore AB, AlphaMOS, and General Electric Life Sciences. In the years 1993–2012 he was a scientific advisor in the Warsaw office of GE Healthcare (formerly Amersham Bioscience, Pharmacia, and LKB). In 1994 he obtained his habilitation in medical biology, and in 2003 the title of Professor of Medical Sciences in the field of medical biology. In 2000 he founded the Department of Biophysics at the Lodz University of Technology which he managed until the end of 2021, coordinated the Center for Advanced Technologies BioTechMed, and participated in the implementation of the BioNanoPark project at the Technopark in Lodz, Poland, where he was employed as the scientific leader of Bionanopark Laboratories (2012–21). From 2012–20 he represented Poland in the Nanomaterials Expert Group of the European Chemicals Agency, Helsinki, Finland. Currently, he is a Professor at the Institute of Materials Science and Engineering of the Lodz University of Technology.

Magdalena Walkowiak-Przybyło

Magdalena Walkowiak-Przybyło, a graduate of Pharmacy at the Medical University of Lodz, Poland (2009), obtained her PhD in materials science at the Lodz University of Technology, Poland (2014). During her PhD studies (2009–14) she was granted several scholarships (e.g. 'Scholarships to support innovative research of PhD students' (2009), 'Research scholarships for the best PhD students in the field of new technologies' (2012–13)). From 2019 to 2021 she worked as a Scientific and Technical Specialist in the field of materiomics at the Institute of Materials Science and Engineering, Lodz University of Technology, and in the years 2020–21 she was employed as an Assistant Professor at the same institute. During the years 2009–20, in parallel with her scientific path, she continued working as a pharmacist in a community pharmacy. Currently she works for a pharmaceutical company, involved in medicinal product development in foreign markets with a great emphasis on regulatory activities.

Piotr Komorowski

Piotr Komorowski graduated from the Lodz University of Technology, Poland, in 2003 with a Master's Degree in Mechanical Engineering, specialty—Medical Apparatus and Equipment. He started his scientific career at the Department of Biophysics of the Institute of Materials Science and Engineering at the Lodz University of Technology. After receiving his PhD in materials science (2010), he took the position of Head of the Laboratory of Molecular and Nanostructural Biophysics and the Laboratory of Structural Research of Nanomaterials in BioNanoPark Ltd, Lodz, Poland. Dr Komorowski has completed several training courses in molecular biology, proteomics, and biophysics, and is a specialist in the field of biological evaluation of nanomaterials and other products of innovative technologies. Has extensive experience in implementing the current techniques used in cellular and molecular biology, including cell viability, cytotoxicity and genotoxicity, as well as transcriptomics and proteomics.

Symbols

2D-DIGE	two-dimensional differential gel electrophoresis
AAS	atomic adsorption spectrometry
AC	affinity chromatography
Ac(Y–Z)	acceptor represented by Y atom linked covalently with the Z atom
AES	atomic emission spectroscopy
AFM	atomic force microscopy
AFS	atomic force spectroscopy
BET	adsorption of inert gas method
c	the speed of light in a vacuum (299 792 458 m s^{-1})
CCM	constant current mode
CGM	constant gap width mode
CHM	constant height mode
CI	chemical ionization
CJD	Creutzfeldt–Jakob disease
CRISPR	clustered regularly interspaced short palindromic repeats
DDT	dichlorodiphenyltrichloroethane
DFT	density functional theory
DLS	dynamic light scattering
DNA	deoxyribonucleic acid
E	energy (J, eV)
E_A	energy of reaction activation
E_n	self-energy of an electron in state n
EA	chemical composition analysis
ELI	Ti6Al4V sample produced by CNC machining
ELI-EBT	Ti6Al4V sample produced by electron beam technology
ELISA	enzyme immunoassay
EU FP7	Seventh Framework Programme of the European Union
FFF	fractionation in laminar flow
EPR	electron paramagnetic resonance
ESI	electrospray ionization
FC	fold change
FDI	field desorption and ionization
GFC	gel filtration chromatography
$\hat{H}$	Hamilton operator in differential form
h	Planck constant (6.626×10^{-34} J s)
$\hbar$	Dirac constant $h/2\pi$ (2.109×10^{-34} J s)
HIC	hydrophobic interaction chromatography
IEC	exchange chromatography
IEF	isoelectric electrofocusing
IFA	immunofluorescence assay
i	the imaginary unit of a complex number
IT	ionic trap
ITC	isothermal calorimetry
ITP	isotachophoresis
k	reaction constant
L	liter

LA	light absorption
LFM	lateral force microscopy
m	mass (kg)
MALDI	matrix assisted laser desorption
MES	magnetic and electric separators
MGI	Materials Genome Initiative
MHET	mono(2-hydroxyethyl)terephthalic acid
mRNA	messenger ribonucleic acid
MS	mass spectrometry
MWCNT	multi-walled carbon nanotubes
NMR	nuclear magnetic resonance
$p(\mathbf{r},t)$	probability of finding a particle
PAMAM	poly(amidoamine) dendrimers
PCR	polymerase chain reaction
PE	photoelectron energy
PET	poly(ethylene terephthalate)
pH	concentration of hydrogen ions in logarithmic scale
pI	isoelectrofocusing point
PE	polyethylene
PES	photoelectron spectroscopy
PS	polystyrene
QA	quadrupole analyzer
QSAR	quantitative structure–activity relationship
$\mathbf{r}$	particle position vector
r	reaction rate
R	universal gas constant equal 8.31 J mol^{-1} K^{-1}.
RGDS	peptide Arg-Gly-Asp-Ser
RIA	radioimmunoassay
RNA	ribonucleic acid
RPC	reverse phase chromatography
SAXS	low angle scattering
SEM	scanning electron microscopy
SEC	size exclusion liquid chromatography
SINS	secondary ion mass spectrometry
SNP	silver nanoparticles
snRNA	small nuclear ribonucleic acid
SPM	scanning probe microscopy
STM	scanning tuneling microscopy
STS	scanning tunneling spectroscopy mode
t	time (s)
T	temperature on the absolute scale
TEM	transmission electron microscopy
Ti6Al7Nb	titanium alloy containing aluminum—6% and nobium—7%
Ti6Al4V	titanium alloy containing aluminum—6% and vanadium—4%
TOF	time of flight
TSI	thermospray and ionization
$V(\mathbf{r},t)$	a function describing the value of potential energy at the point described by $\mathbf{r}$ and *t*
XPS	x-ray excited photoelectron spectroscopy

(X)Dn	hydrogen bond donor represented by X atom
ZE	zone electrophoresis
$\Psi(\mathbf{r},t)$	particle wave function
ν	frequency (s^{-1})

IOP Publishing

Biological Evaluation of Materials
The interaction of materials with their environment
Bogdan Walkowiak, Magdalena Walkowiak-Przybyło and Piotr Komorowski

Chapter 1

Introduction

This chapter is devoted to the presentation of our perception of the evolution of the Universe and the related evolution of human thought, leading to the birth of modern physics, as well as life sciences in their present form and their contribution to understanding the modern world.

1.1 What is this book about?

Briefly, this book is about the relationship between materials science and the life sciences, their interpenetration and mutual benefits, quantum effects and their importance on the nanometric scale, and also about the resulting benefits and threats for the ecosphere and for technological progress. It is no secret that we are currently experiencing the most intense development of technologies in the fields of information technology, materials engineering, medical engineering, and biotechnology and clinical practice. Many researchers see these areas as separate, using the traditional methodological approaches and research strategies assigned years ago to the domains they represent. This approach is the result of traditions and habits passed down from generation to generation in traditional schools of science. However, such an approach is far from the interdisciplinary approach that would allow for jointly solving the most serious scientific and technological problems of our age. By carefully analyzing the literature, it is possible to observe an unhurried evolution which, with the passage of time and as a result of new ideas and new technological possibilities, leads to an increasing number of supporters of interdisciplinarity. There is a kind of analogy between the evolution of the Universe and the evolution of scientific and technological thought.

1.2 The evolution of the Universe

As an introduction to this study, let us try to present this analogy. Although the scientifically documented history of the Universe as described in the standard cosmological model goes back almost 14 billion years [1], and recent reports of a

doi:10.1088/978-0-7503-2656-8ch1

new estimate of the Hubble constant reduce this time by about 1 billion years [2], it will take much less time in our study and this history will be compared with the evolution of scientific and technological thought. Until the beginning of the twentieth century, it was widely believed that the Universe had always existed, suggesting that it did not evolve over time. A similarly convenient conservative argument in scientific discussion is that it has always been so, and that it was enough, and that it was good. In 1912 Vesto Melvin Slipher (1875–1969) observed an infrared shift in the spectrum of electromagnetic radiation from 'spiral nebulae', and in 1922 Alexander Friedman (1888–1925) came to a conclusion about the expansion of the Universe from the equations of general relativity. In 1927 Georges Lemaitre (1894–1966) stated in his reflections that the existence of the expansion of the Universe implies the existence of a starting point at which the entire Universe was condensed. This view was not appreciated by contemporary physicists who believed that Lemaitre, as a clergyman, presented his religious views on the creation of the world by God. The work of Edwin Hubble (1889–1953), taking into account the observations made by Milton Humason (1891–1972) published in 1929 [3], describing the relationship between the distances and speed of galaxies, contributed to the acceptance of the expansion of the Universe, which was clearly associated with the existence of a beginning. This is how the Big Bang theory was born. The Universe in its early stages of development was probably filled with extremely high-density energy. As time passed and the Universe expanded, the energy density decreased and successive phase transitions took place, resulting in a quark–gluon plasma in thermal equilibrium with electromagnetic radiation. As a result of the further expansion of the Universe and the lowering of the energy density, the quark–gluon plasma became the source of the baryons (protons and neutrons) formed. From the moment of the appearance of the quark–gluon plasma (about 10^{-35} of a second after the Big Bang), we can talk about the existence of matter endowed with a resting mass. Along with the expansion of space and the decrease in energy density, the value of the equilibrium temperature of the components of the Universe, including the temperature of electromagnetic radiation, decreased. Accordingly, the modern Universe should be filled with isotropic electromagnetic radiation with an equilibrium temperature just below 3 K, corresponding to radiation in the microwave range.

The confirmation of the existence of such radiation, called relict radiation or microwave background radiation, is one of the most important pieces of evidence for the validity of the Big Bang theory. In order not to burden the reader with too many details in the field of the Big Bang theory, we refer interested readers to a popular science book by Steven Weinberg, titled *The First Three Minutes: A Modern View of the Origin of the Universe* [4].

Let us focus our attention on the diagram in figure 1.1. The starting time corresponds to the point at which the Big Bang occurred at the singularity. So far, no scientific theories have been able to correctly describe this singularity itself and its immediate surroundings. All the stories told begin around 10^{-44}–10^{-37} of a second, when the Universe's temperature was in the range of 10^{32}–10^{28} K, which corresponds to an energy density in the range of 10^{19}–10^{15} GeV. On a linear time scale,

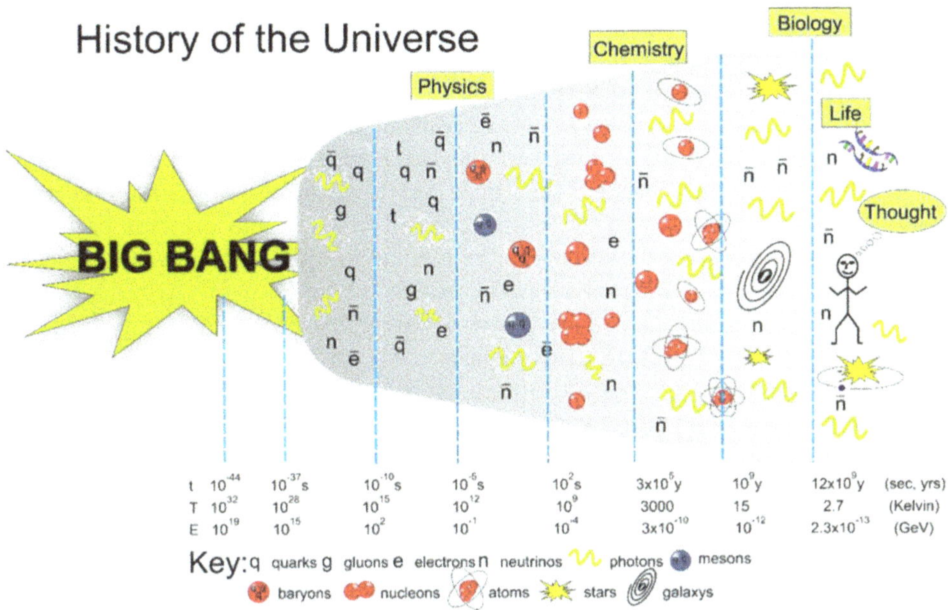

Figure 1.1. Diagram of the evolution of the Universe from its inception (Big Bang) to the present day, indicating the beginnings of the eras of physics, chemistry, biology and the era of human thought.

this interval of events is unimaginably short, but it is very important in its effects. The expansion of the Universe and the related decrease in energy density and temperature allowed the formation of pure energy in space in the form of high-energy electromagnetic radiation and the first elementary particles: quarks and leptons (at a time of about 10^{-10} s). As time progressed, the energy density and temperature decreased, and the successive phase transitions resulted in the appearance of baryons, including protons and neutrons.

This period of time when quarks, leptons, and baryons, as well as other elementary particles, appeared, is often referred to as the era of *physics*. During this period, all transformations were governed only by the laws of physics known to us. This period was much longer and lasted for about 3×10^5 years. At that time the first atomic nuclei (hydrogen and helium) appeared, but the temperature was still too high for the formation of stable atoms containing electrons in addition to the nucleus. At a temperature of about 3×10^3 K the first atoms were formed, thus initiating the era of *chemistry*. This does not mean that the laws of physics have stopped working. The laws of physics have been in force all along with the laws of chemistry. The laws of physics are primary to the laws of chemistry, they imply the laws of chemistry and allow the explanation of chemical observations on their basis. With the further expansion of the Universe and the decrease of the temperature, all antiparticles were annihilated by particles, a small amount of excess particles remained, and the first fluctuations in the density of energy and matter appeared. In places of higher density of matter, this matter condensed and the first stars and star clusters clustered in galaxies appeared. This process came about a billion years after the Big Bang (10^9 years), when the temperature of the Universe dropped below 20 K. Around the emerging stars, planets and other tiny celestial bodies were formed from interstellar matter. It was then that the Earth came to be and captured its satellite Moon from space. The Universe continued to evolve, and after another 2 billion years (about 3×10^9 years after the Big Bang), life appeared on Earth and thus the era of *biology* began. Of course, the laws of physics, and the laws of chemistry resulting from them, were and are still in force, but on their basis a new quality of matter developed, which has the amazing ability to survive through a life cycle, i.e. the ability of self-reproduction, which we take as an attribute of life. It should be borne in mind that all life processes are subject to the laws of physics and the laws of chemistry resulting from physics.

About 525 million years ago, although there are indications even 20–50 million years earlier [5], the first vertebrate animals appeared on Earth, from which primates evolved that gave rise to rational humans (*Homo sapiens*). The beginnings of rational humans date back only 190 000 years [6], which in the time scale of the evolution of the Universe constitutes a very small part, but one very important to us. It was the rational human that evolved towards our contemporary human, creating and eliminating numerous civilizations along the way with all the effects we see today. Without a doubt, we can say that it was only with the existence of rational humans that the thought leading to the emergence and development of science and technology appeared, which we can define as the era of *human thought*.

1.3 The evolution of human thought

At this point, let us return to the previously communicated analogy between the evolution of the Universe and the progress of science and technology. It is not difficult to see that the first human-made stone tools required a certain imagination and precision, but also the use of physical force. This is a very early stage of technological development when other possibilities were unavailable to human thought and experience. With the passage of time and the mastery of using fire, new technological possibilities appeared. It became possible to smelt metals and their alloys, thus entering a new technological reality. It is not difficult to see that there has been a transition from the purely *physical* era to the *chemical* era. Humans began to use a combination of physical and chemical methods of obtaining, processing, and modifying materials in order to produce increasingly complex tools. Since then, physicochemical methods have become the basis of faster and faster technological progress based on rapidly developing knowledge. With time, the rational human noticed that success in hunting and gathering depends heavily on the season, and although this way of living allowed them to survive and have offspring, the domestication of animals and the cultivation of plants allowed for much more comfortable and stable living conditions. Thus, the rational human entered the era of *biology* while creatively developing technologies based on physicochemical knowledge. From then on, physics, chemistry, and biology were helpful in human life, although the humans were not necessarily aware of this. We should not be surprised by the sequence of events in the history of science and the development of scientific thought. Already in the distant past, people observed the relationship between cause and effect and were able to use these observations effectively for their own benefit. This manifested itself mainly in the medical arts, observations of the sky, and the art of war. The accumulated knowledge and experience were passed on from generation to generation by means of mythic messages that easily reached their audience, but often with the defect of a lack of rationalism. The ancient Greeks made a major breakthrough by replacing myths with principles derived from observation through logical reasoning.

The earliest attempts were based on physical observations using simple physical tools to observe the sky and understand the principles of mechanics. This concept has become an integral part of the history of human thought. Originally the sages of that time attempted to describe the common nature of all things called 'physis', but over time there was a division into the material world and into philosophy and theology relating to the spiritual world. An important and long-term effect of the separation of spirit from matter became the interest of thinkers in the spiritual world and the superior role of theology and philosophy over the material world. Documenting the achievements of ancient thought concerning matter we owe a debt to Aristotle (384–322 BCE), who created a coherent, although mostly erroneous, picture of the material world. During the Middle Ages, Aristotle's teachings became part of the Christian doctrine, which effectively froze progress in our understanding the material world for centuries. Only on the threshold of the Enlightenment, thanks to the enlightened minds of Galileo Galilei (1564–1642) and René Descartes (1596–1650), inspired by the astronomical discoveries of Nicolaus Copernicus (1473–1543), was there a change in the perception of the material world. Aristotle's order was replaced with the knowledge of the world through experience (Galileo) and the mathematical description of the world (Descartes). Readers interested in the history of science will certainly be interested in the work titled *A Little History of Science* [7].

For the purposes of this book, we will only recall some facts relevant to the further narrative. The nineteenth century, and in particular the turn of the nineteenth and twentieth centuries, was a period of impressive achievements in human thought, achievements that became the basis of modern science. In the second half of the eighteenth century, in 1763, Luigi Galvani (1737–1798), during the preparation of a frog using metal tools, observed the phenomenon of electrical stimulation of the muscles, which he mistakenly attributed to the so-called 'animal electricity' effect. Despite this incorrect hypothesis, this discovery had enormous consequences, accelerating research into electricity. Soon, in 1800, Alessandro Volta (1745–1827) constructed the first galvanic cell, which became the prototype of all electric batteries and allowed for further research into electricity. In 1808 John Dalton (1766–1844) formulated the hypothesis of the structure of chemical elements from atoms and introduced the concept of atomic weight as a multiple of the weight of a hydrogen atom. The hypothesis assumed that each element consists of indivisible atoms, which are assigned properties characteristic of a given element. In 1815 Jöns Jakob Berzelius (1779–1848) developed a method for determining the atomic mass of chemical elements.

With the discoveries of new chemical elements, there was a need to organize and classify them. Despite numerous attempts, this task remained unsolved for a long period of time, until 1869, when Dmitry Ivanovich Mendeleev (1834–1907) presented his concept of the classification of elements in the form of a periodic table. Mendeleev noticed that the physicochemical properties of elements depend on their atomic weight and that, arranged according to increasing atomic weights, they change their properties regularly with periodic repetitions. The ordering of the elements in the periodic table allowed Mendeleev to indicate empty spaces and approximate the properties of yet unknown elements that were to be discovered in the near future and fill the empty spaces in the periodic table. Mendeleev's achievement is difficult to overestimate—it confirmed the correctness of Dalton's reasoning, narrowed the search area for new elements, allowed the prediction of the properties of substances composed of various elements, and also triggered the acceleration of the development of science, including the science of the atom.

In 1891 George Johnstone Stoney (1826–1911) introduced the concept of the electron as an elementary particle of negative electricity, and in 1897 Joseph John Thomson (1856–1940) experimentally observed the existence of the electron as a particle with a non-zero mass and a negative charge. In 1895 Wilhelm Conrad Röntgen (1845–1923) made another epochal discovery. While working on cathode radiation (the stream of electrons accelerated in a vacuum tube), Röntgen accidentally observed the appearance of radiation much more penetrating than the investigated cathode radiation. This radiation, called x-rays by Röntgen, was introduced into medical practice in 1896 by Ivan Pawlowich Puluj (1845–1918), i.e. much earlier than the mechanism of its formation was explained. Only in the years 1906–1909 did Charles Glover Barkla (1877–1944) discovered the presence of the linear component of x-rays (characteristic radiation of atoms excited by electrons), while the continuous component of x-rays could be investigated only after the formulation of Bragg's law in 1913 (William Henry Bragg (1862–1942) and William Lawrence Bragg (1890–1971)) on the diffraction of x-rays on atoms in a crystal lattice and the construction of a spectrometer enabling the study of x-ray spectra. The success of Dalton's atomic hypothesis opened up the discussion of what an atom is and what its internal structure is like. The first models of the atom appeared and were modified almost from year to year, which shows how intensively the researchers at that time worked on this issue. Generally, the presence of electric charges in an atom was accepted, but the atom was assumed to be electrically neutral, so it had to contain opposite electric charges. In 1901 Jean Perrin (1870–1942) suggested that atoms might look like small planetary systems, a year later, in 1902, Lord Kelvin (William Thomson, 1824–1907) proposed the knot–vortex atom model, which explained the great variety and stability of atoms, and the different possibilities of vibration and spinning of the nodes were to explain the absorption and emission of different light energy at different vibrational frequencies. In 1903 Philipp Lenard (1862–1947), guided by observations in experiments with cathode rays, proposed that every atom consists of empty space and electrically neutral corpuscules called 'dynamids', each consisting of an electron and an equal positive charge. In 1904 Joseph John Thomson, summarizing his earlier proposals, demonstrated the model known as the 'plum pudding'. This model assumed the existence of a sphere with a positive electric charge and electrons contained in this sphere with a negative charge, which was to balance the positive charge of the sphere, so that the entire atom remained electrically neutral. In the same year Hantaro Nagaoka (1865–1950) proposed the first quasi-planetary Saturnian model of the atom. Nagaoka assumed that the atom was made up of a positively charged massive sphere surrounded by numerous negatively charged electrons. The electrons were bound by electrostatic forces, analogous to Saturn's rings, and their oscillations in the plane of the ring emitted radiation with a linear spectrum, while oscillations perpendicular to the plane of the ring gave a spectrum with a band structure. The emission of alpha and beta radiation was the result of the decay of a positively charged massive sphere or ring. In 1910 Arthur Erich Hass (1884–1941) presented his concept of a hydrogen atom in which electronic orbits were quantized, unfortunately the concept initially met with a very critical and even mocking reception, despite the fact that the proposed model allowed the determination of the correct size of the atom. In 1911 Ernest Rutherford (1871–1937) published an analysis of the results of research on the bombardment of thin gold foil with massive alpha particles. The aim of the experiment was to determine the charge distribution in the gold atom. The results of the experiment were in no way compatible with the 'plum pudding' model proposed by J J Thomson, which was accepted at that time. Alpha particles were expected to pass through the gold foil without changing direction or with a slight deviation allowing an assessment of the electric charge distribution in the atom. While it was observed that most of the alpha particles passed through the foil as expected, the trajectories of some of them deviated and some were even bounced back towards the source. This result showed that inside the gold atom there is a concentrated center with a high mass and a positive electric charge, while in the rest of the atom there is a void.

This observation changed the way we think about the atom completely. Rutherford proposed a planetary model of the atom that consists of a small massive positively charged nucleus surrounded by orbiting electrons. The model explained the results of the alpha scattering experiment well, but could not explain the stability of the atom. As a result of the electrostatic interaction, the electrons should fall on the nucleus and the atom should cease to exist. Thus, Hass's proposal from 1910 concerning the quantification of electron orbits returned to favor. The next step was the publication in 1913 by Niels Bohr (1885–1962) of the planetary model of the atom, which combined the Rutherford model with the Hass model. Bohr assumed that electrons can occupy only specific orbits, satisfying the condition that the angular momentum of the electron in this orbit is an integer multiple of the Planck constant divided by 2π ($h/2\pi = \hbar$—the Dirac constant), while the transition from one allowed orbit to the other is associated with the emission or absorption of a portion (quantum) of energy depending on whether the electron goes to a lower or higher orbit. The model was an undisputed success by answering almost all the questions of physicists of the time, including correctly determining the size of an atom, the ionization energy of an atom, and explaining the formation of emission and absorption spectra. It is commonly believed that in 1910 modern physics, in particular quantum physics, was born, based on the simple assumption that energy must be transferred in a portion called a quantum.

1.4 The birth of modern physics

At this point it is worth recalling, at least briefly, the origin and meaning of Planck's constant. Each body at a temperature above absolute zero emits radiation whose total flux is proportional to the body temperature in the fourth power (the law of Josef Stefan (1835–1893) and Ludwig Boltzmann (1844–1906)). It was quickly observed that the emitted radiation is characterized by a continuous spectrum, and the position of the maximum of the emitted energy also depends on the temperature. Lord Rayleigh (John William Stutt 1842–1919) and James Hopwood Jeans (1877–1946) attempted to explain the shape of the emission spectrum by conducting their analyses on the basis of classical thermodynamics. The result of the considerations predicted that the emission capacity of the body should be proportional to the frequency of radiation in the second power. This meant that regardless of temperature, each body should emit the most energy within the range of ultraviolet (high frequency), which was contrary to experience. This discrepancy was called the ultraviolet catastrophe. Max Planck (1858–1947) made an attempt to remove this contradiction and in 1900 he proposed an explanation based on ad hoc assumptions that the particles emitting radiation can be treated as oscillators, the vibrational energy of which can only change in a discrete manner. Therefore they can emit only discrete portions of energy and the size of the energy portion is proportional to the frequency of the emitted radiation, which can be written as the equation:

$$E(\nu) = h\nu \tag{1.1}$$

where E is energy, ν is radiation frequency, and h is Planck's constant.

The adoption of assumptions contrary to the theory in force at that time, according to which the energy of the oscillator can change continuously, allowed a formula correctly describing the shape of the emission spectrum to be derived. It should also be noted here that all these considerations concerned an idealized black-body completely absorbing the radiation incident on it. For a real body, this problem is much more complicated.

A series of experiments confirming the correctness of the assumption about the quantum nature of energy contributed to the emergence of modern physics and its subsequent acceptance. Obviously, the first evidence was the Planck equation, solving the problem of the ultraviolet catastrophe, but the photoelectric effect was already in line, explained in 1905 using the quantum nature of radiation by Albert Einstein (1879–1955), but most of all, the apparent success of the classical Bohr model of the atom contributed to the emergence of quantum physics. In fact, the Bohr model answered almost all the questions that arose at the time, but still failed to answer the question why an atom is stable. The assumption of the existence of permanent stationary electron orbits was contrary to the laws of electromagnetism formulated in 1861 by James Clark Maxwell (1831–1879). According to the theory of electromagnetism, the movement of an electric charge is always accompanied by a magnetic field. If it is a time-varying motion, the magnetic field is also time-varying and creates a time-varying electric field, which in turn generates a time-varying magnetic field, etc. As a result of the time-varying movement of an electron, a time-varying electromagnetic field is created that propagates through space and carries energy. It is not difficult to observe that the electron circulating in the orbit of the atom moves in a time-varying motion (changes its direction), which should cause the emission of an electromagnetic wave, loss of energy by the electron, and its fall into the atomic nucleus. Thus, the atom should not be stable. Experience shows, however, that atoms are relatively stable and certainly are not destroyed by the presence of electrons.

In 1905 Einstein published the special theory of relativity, the crowning theory of classical physics in the field of mechanics, but at the same time building a bridge to modern physics. The special theory of relativity is based on two postulates. The first postulate is the principle of relativity which states that all frames of reference moving in relation to each other in a uniform rectilinear motion (inertial systems) are equivalent. The second postulate speaks of the invariance of the speed of light in a vacuum, which can be put into words that for all inertial observers the speed of light in a vacuum is the same and does not depend on the speed of the system in which the light source is located. These assumptions made it possible to derive a general form of linear transformations of measurable quantities between inertial systems and have a mathematical form of the Lorentz transformation (Hendrik Lorentz, 1853–1928). These transformations contain one invariant parameter of the dimension reciprocal of velocity, which has been interpreted as the reciprocal of the maximum limit speed at which observers can travel, and the maximum speed limit is the speed of light in a vacuum. At low speeds, the correspondence principle is fulfilled and the Lorentz transformation turns into the Galilean transformation, while for speeds comparable to the speed of light relativistic effects appear, such as shortening of the time and distance in the direction of motion, and mass gain for objects with non-zero rest mass. The special theory of relativity implies a number of consequences, the most important of which seem to be the relativity of events, the limitation of the maximum speed, and the conclusion that mass is a form of energy:

$$E = mc^2 \tag{1.2}$$

where E is energy, m is mass, and c is the speed of light in a vacuum.

The special theory of relativity was initially met with disapproval, but contributed to the development of work on relativistic theories of gravity. In 1907 Herman Minkowski (1864–1909) applied the special theory of relativity to space–time and on this basis Einstein developed and published in 1916 the general theory of relativity, which is in fact a relativistic theory of gravity.

Several years had to pass from the formulation of the atomic model by Bohr before the problem of the stability of the atom was solved, and quantum physics gained the recognition of a large group of contemporary researchers. During this period of time, in addition to Einstein's epochal works, very fruitful concepts emerged. Among others, in 1922 Arthur Compton (1892–1962) announced the explanation of the effect of photon scattering on free electric charges (the Compton effect) and in 1923 Louis de Broglie (1892–1987) proposed a hypothesis that matter has a dual particle–wave form. That is, each particle displays its own wave or particle properties depending on the conditions of the experiment, but always only either particle or wave, never both properties at the same time. In fact, this hypothesis provided an alternative description of the particle in relation to the classical description, and at the same time allowed the electromagnetic wave to be treated as a particle (Compton effect—elastic scattering of a photon on a free electron). This way the concept of waves of matter emerged. The de Broglie hypothesis was soon confirmed experimentally by Clinton Joseph Davisson (1881–1958) and Lester Germer (1896–1971), who in 1927 observed electron diffraction on a nickel crystal. The wave properties of the electron were soon used in the construction of the electron microscope, which happened in 1931. Earlier, in 1925 Werner Heisenberg (1901–1976), with significant contributions from Max Born (1882–1970) and Ernst Pascual Jordan (1902–1980), formulated quantum mechanics in matrix representation and one year later, in 1926, Erwin Schrödinger (1887–1961) published a competing quantum mechanics in wave representation. The problem obviously arose as to which representation was correct, but it turned out that both descriptions are equivalent and lead to the same conclusions with regard to the quantum analysis of the atom. However, due to the greater complexity of matrix calculus compared to wave calculus, wave mechanics has become the basis for accepting quantum mechanics and is more commonly used to describe the quantum states of electrons in an atom. The time-dependent Schrödinger equation in a positional representation is

$$\hat{H}\Psi(\mathbf{r}, t) = \mathrm{i}\hbar\frac{\partial\Psi(\mathbf{r}, t)}{\partial t}, \tag{1.3}$$

where

$$\hat{H} = -\frac{\hbar^2}{2m}\left(\frac{\partial^2}{\partial x^2} + \frac{\partial^2}{\partial y^2} + \frac{\partial^2}{\partial z^2}\right) + V(\mathbf{r}, t) \tag{1.4}$$

is a Hamilton operator in differential form, $V(\mathbf{r},t)$ is the potential energy, m is the mass of the particle, $\Psi(\mathbf{r},t)$ is a position and time-dependent wave function, $\mathbf{r}$ is a position vector, i is the imaginary unit, and $\hbar$ is the Dirac constant ($\hbar = h/2\pi$).

This equation is significantly simplified when the analyzed quantum system is isolated from the environment and it can be assumed that the total energy of the system does not change. Then the time-independent Schrödinger equation takes the form:

$$\hat{H}\Psi(\mathbf{r}) = E\Psi(\mathbf{r}), \tag{1.5}$$

where E is the energy of the system.

In 1926 Born proposed a probabilistic interpretation of the wave function from the Schrödinger equation, namely that the square of the module of the wave function is equal to the probability of finding a particle at a point position $\mathbf{r}$ at time t:

$$|\Psi(\mathbf{r}, t)|^2 = p(\mathbf{r}, t), \tag{1.6}$$

where $\Psi(\mathbf{r},t)$ is a wave function depending on the position and time t, while $p(\mathbf{r},t)$ is the probability of finding the particle at the point described by the vector $\mathbf{r}$ and time t.

By solving the stationary Schrödinger equation, one can find many different wave functions $\Psi_n(\mathbf{r})$ satisfying the equation, called eigenfunctions (self-functions), but always in tandem with the corresponding energy E_n, called the eigen-energy (self-energy). Each of the pairs $\Psi_n(\mathbf{r})$ and E_n describes a separate energy state of an electron allowed in an atom in the potential field of an atomic nucleus and other electrons. The index n determines the successive self-energy levels and the corresponding orbitals described by its eigen-wave (self-wave) function. Quantum physics does not determine the shape of the orbit of an electron in an atom, but only indicates the area of space in which an electron can be found with a certain probability.

The transition of an electron from one energy state to another is always associated with a change in its self-energy and its self-wave function. In the case of the spontaneous transition of an electron from a state with higher energy to a state with lower energy, a quantum of energy (photon) is emitted with a strictly defined energy value corresponding to the difference in energy levels:

$$E(\nu) = h\nu = E_n - E_{n-1} \tag{1.7}$$

where ν is the frequency of the photon, and E_n and E_{n-1} are the nth and $(n-1)$th electron energy states allowed in an atom.

It follows that the transition from the state with lower energy E_{n-1} to the state with higher energy E_n requires absorption by the atom of a photon with a strictly defined frequency and energy corresponding to the difference in the values of these energy levels.

Quantum mechanics, regardless of the selected representation, be it matrix or wave, explains the durability of the atom, and the successive achievements of modern physics are based on concepts and laws formulated at the beginning of the twentieth century, which does not mean that science has reached the end of its road. Each new discovery becomes a source of new hypotheses and new laws, and knowledge seems to have no limits. This is illustrated quite well by an anecdote from the late nineteenth century. When Max Planck was considering the choice of the path of further education, the Munich professor of physics, Philipp von Jolly (1809–1884) strongly advised him against choosing physics as a field of study, claiming that virtually everything had already been discovered and there was nothing important left to do. Fortunately, Planck did not follow this advice and gave a very strong impetus to the development of physics, the effects of which we appreciate today.

Also in the field of chemistry related to physics, this period of time had extremely fruitful consequences. The subjects of interest in physics were individual atoms and their internal structure and interaction with radiation, while in chemistry the subjects of interest were atoms and the possibility of their joining into groups of atoms, i.e. the molecules of which the substances surrounding us are composed. On the basis of discoveries in physics, chemistry also developed very dynamically, in many cases supplementing the discoveries of physics or participating in them. In fact, it was difficult at that time to clearly separate physics from chemistry, and the researchers of that time dealt with both physics and chemistry, thus meeting the requirements of the currently observed interdisciplinarity.

Chemistry took a position strongly opposed to alchemy. First of all, chemistry rejected the abstract notion of alchemical elements symbolizing opposing qualities. Elements were considered homogeneous substances obtained as a result of the analysis of complex substances, and Antoine Lavoisier (1743–94) did not prejudge the number of elements, or that substances considered as elements would not be broken down into more elementary substances in the future, i.e. the proper elements. The second fundamental difference was the adoption of the axiom of the invariability of elements, which contradicted the possibility of the transmutation of elements, i.e. the transformation of elements into other elements, which was sought in alchemy in trying to find a philosopher's stone that could change lead into gold.

Substances were divided into two groups depending on their origin, and the concepts of animate matter coming from the living world and other inanimate matter were introduced. The direct result of this was the division of chemistry into two parts, organic and inorganic chemistry, a division that remains valid today. In the field of organic chemistry, there was a place to track changes taking place in living organisms, which was the basis of modern biochemistry. It is worth mentioning a few epochal achievements in the chemistry of those times.

In 1896 Henri Becquerel (1852–1908), while studying the phosphorescence of minerals, discovered the radioactivity of uranium ore. The study of the radioactivity of uranium ores was undertaken by Maria Skłodowska Curie (1867–1934), who showed that the radiation discovered by Becquerel is a property of the uranium atom and also proved the radioactivity of another element—thorium. Her husband Pierre Curie (1859–1906) supported her in her research. In 1898 they announced the discovery of a new radioactive chemical element called polonium, and a few months later another previously unknown element—radium. Their work on isolating the discovered elements in their pure form contributed to the development of instrumental and analytical methods in chemistry, and the discovery of natural radioactivity became the foundation of the physics of the atomic nucleus. The result was the discovery of nuclear reactions and artificial radioactivity. Thus, the alchemists' dream came true, as a result of nuclear reactions, the transmutation of elements was possible, and the transformation of mercury and lead into gold was observed in the laboratory. Unfortunately, at the moment the transmutation technology, however possible, is completely unprofitable.

In 1916 Gilbert Newton Lewis (1875–1946), in 'The atom and the molecule' [8], explained the nature of the chemical bonds that bind atoms into a molecule. This concept is valid to this day and contributed to perhaps the greatest achievement in twentieth century chemistry, which is the production of polymer plastics. The foundations for polymer chemistry were prepared by Hermann Staudinger (1881–1965), who formulated the concept of the macromolecular structure of polymer components, contrary to the beliefs of his contemporaries that there was no possibility for the existence of large macromolecules.

The group of artificial materials includes both those for the production of which natural polymers are used, mainly cellulose, and those which are synthesized from scratch in a technological process in which hydrocarbons are used as starting substances. The mid-twentieth century is the moment when polymer chemistry began its triumphant development. Many different polymer materials have been developed with a very wide spectrum of physicochemical and mechanical properties and durability, often exceeding natural products and having practically unlimited use in all areas of human activity. We will not make a mistake in identifying the second half of the twentieth century as the era of plastics.

The success of polymer plastics has led to their mass production with a shortage of processing technology or controlled degradation. Therefore, the advantage of the very high durability of plastics has become their disadvantage, leading to the accumulation of used plastics in the form of barely degradable waste. Mechanical shredding of plastics does not lead to their chemical or biological degradation, but introduces a risk difficult to assess for living organisms in the aquatic environment, soil, and air. It is also worth noting that the problems with durability and environmental risk from plastics are mainly attributed to packaging, while textile products based on plastics pose a much greater risk by producing micro- and nano-particles during their use; in particular many of these particles are formed during the washing of clothing.

Another spectacular achievement of chemistry is the development of diamond synthesis methods. Unlike in the case of the transformation of lead or mercury into gold by nuclear reaction methods, in the case of diamond synthesis one does not have to penetrate into the atom, but only provide the appropriate conditions for the synthesis process. The first synthetic diamonds with the possibility of production on an industrial scale were obtained in 1955 in the laboratory of General Electric (GE) in the USA in a high-pressure–high temperature (HPHT) process mimicking the natural conditions of diamond synthesis. Even before the success of GE, in 1954 a much milder diamond production technology was patented, consisting of the low-pressure and low-temperature synthesis of a diamond crystal from the gas phase using the chemical vapor deposition (CVD) method. However, due to problems with the repeatability of this method, the industrial implementation took place only in the 1980s. Synthetic diamonds, despite their physical and chemical properties being identical to natural diamonds, were not used in the jewelry industry due to their relatively small size. Diamond powders, on the other hand, have found wide industrial use as abrasive or cutting materials. Detonation methods and plasma assisted vapor deposition (PAVD) methods are often used to produce nanodiamond powders. The positive feature of this achievement in relation to the success of polymer chemistry is the absence of environmental hazards from synthetic diamonds.

1.5 Life sciences and their contribution to understanding the world

Physicochemical methods of getting to know the material world have become a permanent part of science. Modern materials science in particular is based on the knowledge and experience of generations of researchers. For a long time, representatives of material science have been convinced, and many of them still believe, that the physicochemical methods of producing and assessing the properties of manufactured materials are good enough to also assess their safe use. Unfortunately, the effects of treating materials science in this way has brought us to today's very unfavorable situation of pollution of the environment in which we live, with all its negative consequences.

It seems that the inclusion of the achievements of biology and life sciences into the field of materials science is now becoming a necessary action to attempt to solve the problem of the safe use of modern material technology products, reduce environmental contamination, and effectively remediate it. In short, the age of biology in materials science has arrived, and it is complementary to the era of physics and chemistry. The advent of the era of biology in the development of the Universe allowed for the emergence of the rational human and the development of intelligent thought. In analogy, with the era of biology in materials science there is the possibility of solving the most complex problems of materials science and intelligent materials technology. This is largely due to the observation of material and construction solutions that nature has developed in the course of the evolution of living organisms. Picking these solutions for their technological use, referred to as bionics or biomimetics, resulted in ideas for the production of previously unavailable materials, often hybrid, and the use of these materials for the construction and production of previously unavailable products.

Modeling on solutions developed by nature also indicates thus far unknown possibilities for managing society and the economy. The cell seems to be an unsurpassed example of effective production management, and the entire body can be used as a model solution for managing a community of cells that create the tissues and organs necessary for the functioning of the body. It is enough to mention that in a eukaryotic cell, several thousand genes are simultaneously expressed, which translate into the production of an assortment of several thousand gene expression products—peptides and proteins. This production process is managed in such a way that it takes place without unnecessary stocks of substrates and without storing products and, if necessary, cells can be divided, thus increasing the number of cells ready for increasing the production of proteins. Both the substrates and products are delivered on an ongoing basis, depending on the needs. If there are production errors in the cell, usually resulting from errors (mutations) in the genetic material, the cell activates the repair apparatus and returns to its normal functions. However, if the errors are too extensive and their repair is impossible, the cell initiates the process of programmed death—apoptosis. In the language of economics, this is the collapse of a business—bankruptcy. Of course, it also happen, although very rarely, that a cell escapes programmed death and becomes a cancer cell with the features of immortality. In the vast majority of cases, such an immortalized cell is detected by the body's immune system and is eliminated. It turns out, however, that nothing is perfect and the body's natural regulatory and defense mechanisms sometimes fail, the immune system is unable to properly recognize a cancer cell that has developed a strategy to hide its cancer features from recognition, which leads to the development of cancer.

However, it should be noted that this dramatic event for the body, which is the onset of a cancer as a result of a mutation of genetic material, occurs with an extremely low probability, considering that the average adult human body consists of about 4×10^{13} cells [9], of which each is involved in the current expression of several thousand genes necessary for proper functioning, and the lifetime risk of cancer does not exceed 45% [10]. The appearance of cancer in the language of economics means a hidden deviation from the principles of the market economy. The appearance of the first signs of an economic crisis, in the absence of appropriate economic and financial therapy, may lead to the collapse of the national economy or even the global economy. Economic crises occur in the scale of human life much more often than cancer diseases.

The management mechanisms of a single cell, as well as a set of cells that make up an organism, are also worth getting to know in detail because of the way information is stored and transmitted. In the distant past, alchemists searched for a philosopher's stone that would enable the transmutation of common metals into precious metals. This idea has never been realized (except for the changes of nuclei as a result of nuclear reactions, mentioned earlier in this chapter), but the idea of obtaining a universal material has never been abandoned. To this day we use the concept of the versatility of various materials for specific applications. In this respect, the materials most frequently called universal materials are polymeric materials, in particular those used in optoelectronics [11, 12]. New universal materials with intentionally defined properties and versatile applications are still a dream to come true. The current state of knowledge and technological advancement do not allow the production of a 'super material' suitable for every application. Viewing the solutions developed by nature allows you to identify a strategy that may help solve this problem in the future. In the case of cells, tissues, and whole multicellular organisms, nature has found an ingenious solution. All polymeric materials present in cells of various tissues, characterized by different mechanical and functional properties, are synthesized based on the genetic information contained in the cell nucleus. Regardless of the role and function of a cell in a given tissue, the genetic material in each cell is identical and contains the same information. However, depending on the role and function of the tissue, the cells that compose it use only a fraction of the information concerning their functioning. In this sense, the universal 'super material' is deoxyribonucleic acid (DNA), which contains information about the functionality of the polymer molecules synthesized in the cell. Perhaps this is where the path leads to solving the problem of the technology of producing universal materials based on the technology of producing and using the superior 'super material'.

As a result, the knowledge and technologies at work in the field of life sciences can become very useful in the field of materials science and materials manufacturing technologies. It would also be impossible to advance in the field of life sciences without the revolutionary solutions provided by materials science and materials engineering and applied in advanced research devices. Currently, both of these areas merge with mutual benefit using the knowledge of physics, chemistry, and biology. A good illustration of this interaction can be a puzzle pattern, which creates a new, better quality form under the conditions of a snug fit (figure 1.2). The best example of this is the emergence of a new strategy in the field of materials science at the beginning of this century, a strategy based on proven strategies in the area of life sciences. We are talking about *materiomics*.

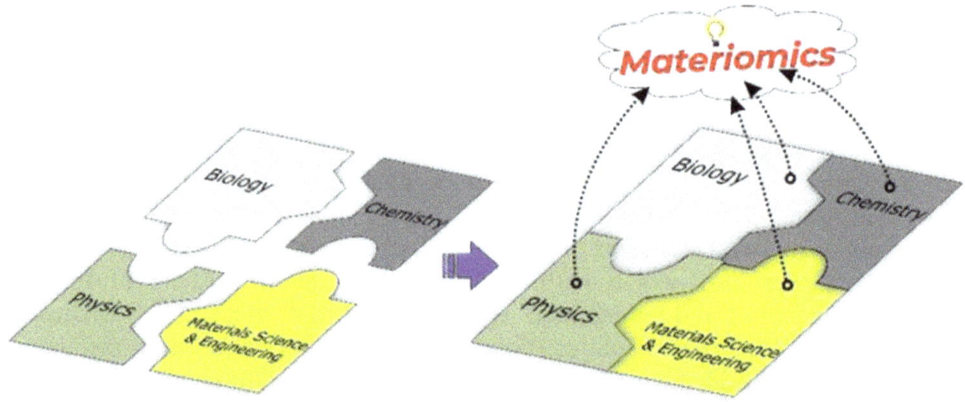

Figure 1.2. Schematic of the cooperation of basic sciences within the framework of the modern understanding of the structure of materials science covered by the materiomics strategy.

Although material engineering is often, but completely wrongly, associated by the uninitiated only with tailoring and textile materials, materials science, of which material engineering is an integral technological part, is the most interdisciplinary science, basing its research on traditional methods from physics and chemistry, but also more and more often on methods derived from biology and information technology. Materials science allows you to check your idea of the material world without distinguishing between animate and inanimate matter and without differences in the size scale of material objects. The laws of physics are the same for the entire material world, both macroscopic and microscopic, and therefore they also determine the biological reality, which is extremely rich and complex, and therefore an extremely interesting challenge for the researcher.

It is also worth paying attention to the fact that during the heated discussion on the ecological effects of the production of modern materials, including nanomaterials, and their modification and further processing, biotechnological methods are becoming an underestimated source of inspiration and ecologically clean material technologies. Modern research in the field of materials science also uses information technology and opportunities defined by life, health, and nutrition sciences, which further deepen its interdisciplinary character. Materials science, incorporating or working closely with biotechnology, will develop extremely rapidly, comparable only to information technology and the sciences of life, health, and nutrition. For these reasons, it is worth participating in research in the field of materials science.

References

[1] Ade P A R *et al* 2016 Planck 2015 results. XIII. Cosmological parameters *Astron. Astrophys.* **594** A13

[2] Schombert J, McGaugh S and Lelli F 2020 Using the baryonic Tully–Fisher relation to measure H_o *Astron. J.* **160** 71

[3] Hubble E 1929 A relation between distance and radial velocity among extra-galactic nebulae *Proc. Natl Acad. Sci.* **15** 168–73
[4] Weinberg S 1997 *The First Three Minutes: A Modern View of the Origin of the Universe* 1st edn (New York: Basic Books)
[5] Shu D-G *et al* 1999 Lower Cambrian vertebrates from south China *Nature* **402** 42–6
[6] Hammond A S, Royer D F and Fleagle J G 2017 The Omo–Kibish I pelvis *J. Hum. Evol.* **108** 199–219
[7] Bynum W 2012 *A Little History of Science* (London: Yale University Press)
[8] Lewis G N 1916 The atom and the molecule *J. Am. Chem. Soc.* **38** 762–85
[9] Bianconi E *et al* 2013 An estimation of the number of cells in the human body *J. Ann. Hum. Biol.* **40** 463–71
[10] White M C, Holman D M, Boehm J E, Peipins L A, Grossman M and Henley S J 2014 Age and cancer risk: a potentially modifiable relationship *Am. J. Prev. Med.* **46** S7–15
[11] Qiu X *et al* 2020 Universal host materials based on carbazole-formate derivatives for blue, green and red phosphorescent organic light-emitting diodes *Dyes Pigm.* **174** 108045
[12] Wu C, Wang B, Wang Y, Hu J, Jiang J, Ma D and Wang Q 2019 A universal host material with a simple structure for monochrome and white phosphorescent/TADF OLEDs *J. Mater. Chem.* C **7** 558–66

IOP Publishing

Biological Evaluation of Materials
The interaction of materials with their environment
Bogdan Walkowiak, Magdalena Walkowiak-Przybyło and Piotr Komorowski

Chapter 2

Materials science and life sciences

The interaction of materials science and the life sciences allows us to understand the impact of the production and use of the products of materials technology, considered against the background of the industrial revolution, on the state of the environment. The potential of micro-organisms to decompose waste and participate in technological processes, as well as problems related to risks from nanotechnology products are considered.

2.1 The impact of the production and use of material technology products

For a long time, the representatives of materials engineering were convinced, and many of them still believe, that the physicochemical methods of production and evaluation of the properties of manufactured materials are also good enough for the evaluation of their safe use. It is extremely rare to carefully consider the threat posed to man and the environment by the process of degradation of manufactured materials. At most, the problem to be solved was to reduce the effects of corrosion and material consumption in order to extend the life of the finished products, but this problem also ceased to be important over time, because the modern production and sales philosophy clearly assumes that the finished product is to only function reliably during the warranty period. After this period, the product should be withdrawn from use and replaced with a new one.

This line of reasoning drives production and brings specific benefits to the producer and seller. However, this is not necessarily beneficial to society as a whole and to the environment in which humans live. The fact is that industrial products during their use and after their end of life, unless properly managed and processed, are a source of degradation products. Therefore, in a circular economy, an important role is assigned to the selective collection of waste and its processing.

The causes of degradation are obviously physical factors, which are responsible for mechanical wear, chemical factors, which are responsible for corrosion, and

doi:10.1088/978-0-7503-2656-8ch2

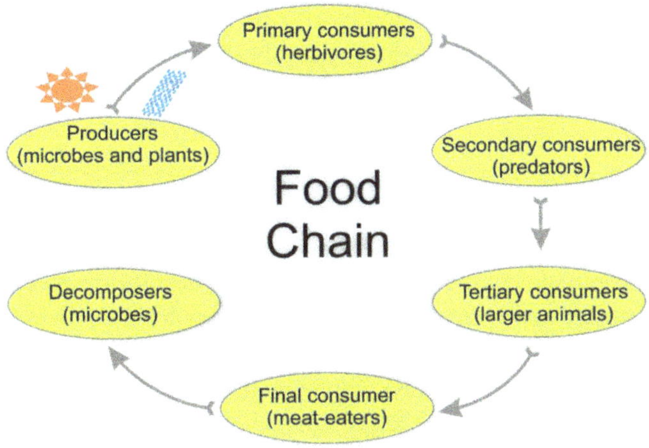

Figure 2.1. Food chain diagram showing possible routes for permanent degradation products reaching our bodies.

biological factors, which are responsible for biocorrosion. Moreover, the products of degradation by wear or corrosion (regardless of the mechanisms) provide fine particles of material and the resulting chemical compounds that accumulate in the environment and become part of the food of micro-organisms, plants, and simple animals (the primary consumers). In this way, they enter the food chain and reach our bodies unnoticed (figure 2.1).

2.2 The industrial revolution and its impact on the environment

We learned the importance of the problem using the example of plastics, which, both in the form of whole objects and small fragments resulting from mechanical or chemical degradation, have polluted the waters of the seas and oceans to such an extent that significant amounts are also found in the seas of the Arctic and Antarctic regions [1, 2] which were previously considered pure. On an unknown scale, this problem also applies to plastics accumulated on the surface of continents, where the degradation products of these plastics freely migrate with rainwater, rivers, and groundwater [3–5], and have already reached our bodies. The management of metal products, which are readily recovered for processing, looks much better. But, here too, a significant amount of used metals is not recycled, and the corrosive processes supply the ecosystem with significant amounts of metal particles and their ions. A separate group of degradation products are the products of innovative surface engineering technologies. These technologies make it possible to obtain a wide variety of surface structures with very diverse chemical compositions and properties. Surfaces modified in such a way also degrade and the environment receives particles and chemical compounds with undefined potential for threats to living organisms. This is a previously unknown situation for the ecosystem, just as the common presence of plastics and their particles is new, which makes it impossible to fully assess the risk posed by these materials. The importance of the problem is reflected in the priorities of the European Commission [6], in particular the elimination of pollution, the maintenance of biodiversity, the use of clean forms of energy, and measures to improve the climate. This strategy includes the elimination of fossil fuels, the extension of the lifetime of technological products, and their reuse. Considerations regarding the risks posed by used products were also presented by the authors in the local journal *Laboratorium—Przegląd Ogólnopolski* (*Laboratory—Polish Nationwide Review*) [7].

We need to realize that in the past, technological progress was relatively slow, and therefore living organisms had enough time to start adaptation processes that would enable proper functioning under new conditions, as long as the concentration of degradation products in the environment did not exceed the critical values for adaptive capacity. We pride ourselves on the dynamic development of an economy dependent on technological progress. The first stage of the industrial revolution came at the end of the eighteenth century, when the steam engine appeared and the mechanization of production began. The second stage began almost a hundred years later (at the end of the nineteenth century), when electricity and mass production were used in industry under the conditions of specialized operations on the production line. The third stage is identified with the introduction of electronics and automation in the 1970s, while the current fourth stage began at the end of the twentieth century and is characterized by widespread access to the Internet, the development of genetic engineering, the mastery of 3D printing, and the globalization of the economy [8]. Without a shadow of a doubt, none of the next stages of the industrial revolution could have happened without the significant contribution of materials science to technological development.

Not only is the time between particular breakthrough moments shorter, at the same time, despite numerous disturbances, the gross national product has grown strongly non-linearly during this time [9], which may be a measure of the growth in industrial production. It is expected that we will soon (at least until the middle of the twenty-first century) move to the next, fifth stage of the industrial revolution, characterized by the large-scale use of bio- and nanotechnology in conjunction with artificial intelligence and in close co-operation between humans and intelligent machines. It will be related to changes in social relations aimed at reducing social inequalities through universal access to modern technologies and with the production of high-quality products adapted to the needs of individual people (customized products) (figure 2.2).

This positive image, resulting from the development of knowledge and technological progress, responsible for the next stages of the industrial revolution, is unfortunately greatly disturbed by the unexpected and unwanted effects of this revolution on the human environment. We would like to point out here the possibility that the same industrial revolution gives us to monitor the threats and changes already made in the environment, as well as the potential possibilities of counteracting these threats and remediation (repair) or more broadly reclamation (regeneration) of our environment. To activate these opportunities, first of all, the will of decision-makers is needed to undertake such actions quickly and on an appropriate scale.

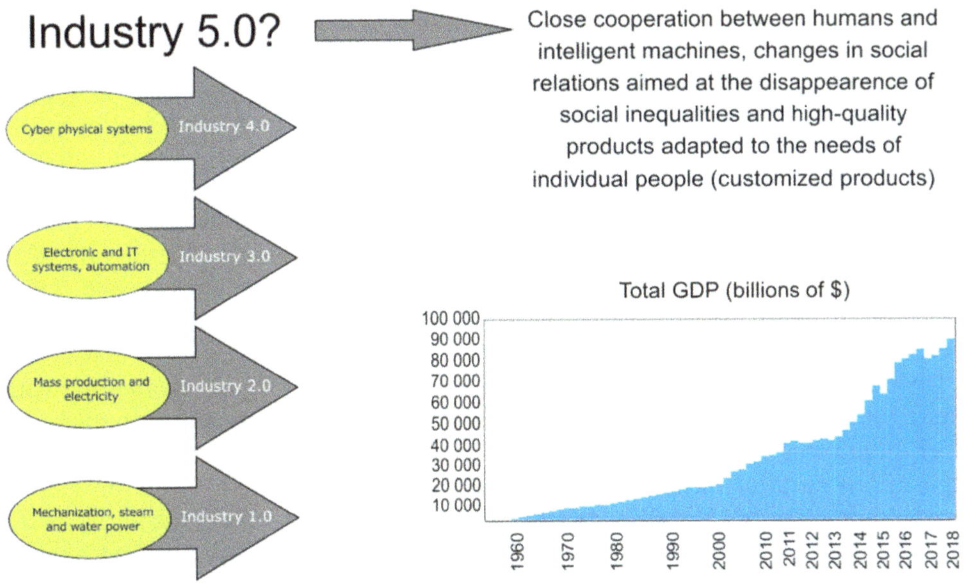

Figure 2.2. Diagram of the stages of the industrial revolution with the presentation of total gross national income.

2.3 Environmental monitoring

The simplest solution seems to be to track, with the help of measuring equipment, the presence of pollutants resulting from technological human activity. Unfortunately, such a solution turns out to be expensive and has very limited spatial possibilities. Organisms living in soil, water, and air come to the rescue. These organisms feed on simple organic substances available in their environment, particularly in soil and water, but also decompose complex organic residues. We can successfully treat these organisms as biological indicators of environmental quality, and at the same time we can see them as biological systems capable of removing pollutants and regenerating the environment.

Let's start with the surface of the Earth, where huge amounts of both municipal and industrial waste are stored. The problem of the persistence and decomposition of the solid waste collected on the Earth's surface is very complicated, and the observed processes involve physicochemical and biological factors. Shredding solid waste, through its mechanical impact and corrosion (chemical and biological), leads to the release of various substances which, according to the laws of physics, migrate in the soil environment, reach the groundwater and river waters, and appear in the air in the form of gases and dust suspensions. Living organisms play a very important role in the degradation processes of solid waste, ranging from simple prokaryotic organisms, through simple plants and fungi, to consortia of micro-organisms and multicellular organisms. We encourage the reader to deepen their knowledge in this field and refer to a very accessible monograph *Chemistry and*

Biology of Solid Waste [10]. Of course, liquid waste is also an environmental problem, the more so as a significant part of it still does not end up in wastewater treatment plants, particularly in developing countries [11]. A separate part of the natural environment, although closely related to soil and water, is air. In the air, you can find both small molecules of chemical compounds in the form of gases and much larger airborne particles in the form of relatively persistent suspensions (fog, smog, smoke, viruses, micro-organisms, and many others). In particular urban areas are exposed to the formation and presence of such suspensions [12]. The quality of the environment and its impact on our lives depend highly on the amount and type of substances accumulated in the soil, water, and air, and on how exposed we are to contact with these substances.

Living organisms can successfully act as risk indicators. Organisms that we would like to use for these purposes must be more sensitive to environmental changes than other organisms living in this environment, must be strong enough to survive exposure to harmful factors, and must be particularly sensitive to the presence of certain substances in the monitored environment. In order to fully assess the body's response to stress caused by contact with harmful factors, many types of observations should be made, including population observations (viability and reproductive potential), morphological (size and morphological features), and vital activity (behavioral changes) in terms of predatory traits and competition for living space. Both invertebrate organisms, such as insects (e.g. mayflies, stoneflies, and butterflies) or worm-like invertebrates (earthworms and nematodes), and vertebrate organisms (mainly fish and birds, but also small mammals) can be used successfully. It is worth noting that the quality of the environment can also be demonstrated by the population, health, and behavior of large wild animals living in natural conditions, and of plants that contribute to these natural conditions.

As a point of curiosity, but also as an example of the practical application of bioindicators, it is worth citing the technology used since 2019 in the water intake for the city of Warsaw. Warsaw is supplied with drinking water from an intake under the bottom of the Vistula River (the Gruba Kaśka well) and from Lake Zegrzyński [13]. Long-lived mussels of the species *Unio tumidus*, very sensitive to water pollution, were used to monitor the water quality on a current basis. These mussels live for several dozen years under natural conditions, but after several months of work in the Warsaw Waterworks, they are returned to their natural habitat. After catching the mussels, they undergo a period of acclimatization and are equipped with sensors the size of the opening of the shells, and then they are placed in a tank with the flowing collected water. These sensors are connected to the water intake control system. Clams eat the organic matter contained in the water and their shells remain open. When the water quality deteriorates, the mussels close their shells to isolate themselves from the environment, and the opening sensors transmit this information to the system and cause the water intake to close immediately. It is not only the mussels that watch over the quality of the water in the Warsaw Waterworks. Freshwater fish are also employed for this purpose, such as certes, barbels, and rudds. The fish are placed in specially adapted flow-through aquariums filled with flowing water and their behavior is constantly monitored by a system of cameras transmitting the images in real time to the plant's control room. Any unusual behavior of the fish is recorded and signals the initiation of appropriate laboratory tests.

The quality of the soil can be successfully monitored with the use of earthworms, which, in favorable conditions of humidity and the availability of organic matter and with limited salinity, significantly increase their population and, thanks to their high activity, enrich the soil with products of their metabolism [14]. The presence of earthworms and their high activity is an indicator of good soil conditions. Apart from earthworms, the presence and undisturbed activity of ants and termites may indicate good conditions and high-quality soil [15]. On the other hand, the presence of certain micro-organisms in the soil indicates its degradation and contamination as a result of human activity. In this way, it is possible to demonstrate the presence of human feces and household waste as well as industrial pollution with heavy metals, oils, and other technological waste [16].

Air quality is mainly reflected in the vegetation of plants, while mosses and lichens, due to their poorly developed root systems, are a very convenient indicator of the presence of pollutants in the air and in the top layers of soil [17]. Trees and forest complexes constitute huge biological systems that are extremely sensitive to air pollution and able to actively influence its quality by removing both simple chemical compounds and particles suspended in the air [18]. Birds, whose behavior, population, and activity depend strongly on the presence of pollutants in the air, are also an invaluable indicators of air quality [19]. The quality of soil, water, and air is highly interdependent and should not be considered separately. Changes in the aquatic environment have an impact on soil and air, and in this convention, signals from all bioindicators should be analyzed for information on the quality of the environment.

The above-mentioned problems with the uncontrolled increase in the amount of plastics produced and their accumulation in the environment require a very urgent solution. The mere selective collection of municipal and industrial waste, without its further processing into useful forms, leads nowhere. It becomes necessary to design such plastics that can easily undergo a controlled process of biodegradation to simple carbon compounds that occur in nature and do not pose a threat to living organisms and that can again be a substrate for industrial processes.

We deliberately mention biodegradation here and avoid mechanical and chemical degradation, which can possibly only be used as an aid or in the process of transforming waste into a useful product. The reason for this is that the mechanical fragmentation of plastics does not solve the problem of waste, although it removes macroscopic waste being in front of our eyes. The particles created as a result of this process end up in the environment and, as a result, also in the food chain, the final link of which is humans. Chemical degradation additionally involves large amounts of chemical reagents, which cannot always be used in a closed cycle and, unfortunately, most often also end up in the environment. It is worth noting that some plastic waste has been used for the production of new products for a long time. An example is the well-known and valued textile product commonly known as fleece, which is produced from PET packaging (e.g. bottles). The problem, however, is the extremely durable non-woven fabric, from which micro- and nano-fibers are released during use, and at the end of use, the problem of another form of municipal waste remains. Moreover, the need for fleece is limited, while the amount of PET material produced is far beyond its ability to be processed in this way. A similarly durable material is polystyrene (PS), from which Styrofoam is made, among others, widely used due to its very good thermal insulation properties. Another durable and commonly used material is polyethylene (PE), which, due to its high degree of chemical inertia, is commonly used as packaging for food products (foils, shopping bags, bottle caps), construction elements (pipes, cable insulation, construction elements), and in medicine (implants). Polyethylene, like PET, is used to re-manufacture products, but here too the amount of polyethylene produced exceeds the recyclability. Of course, these are not the only polymer plastics produced on an industrial scale and accumulated in the environment in large amounts. The list of these materials is long and continues to grow due to the introduction of new or modified materials. It is at the stage of developing these new or modified plastics that it would be worth considering their future effective biodegradation.

2.4 The potential of micro-organisms to decompose waste

In some ways, nature itself has taken the fate of the environment into its own hands. It turns out that micro-organisms capable of degrading individual materials are appearing more and more often, using degradation products as a source of the carbon necessary for life. Already in 2008 [20], the existence of fungi producing the enzyme digesting PET (polyethylene terephthalate-hydrolyzing enzyme) was reported. The intended application of this enzyme was the surface modification of fibers made of this material.

It soon turned out that bacteria were also capable of producing PET digesting enzymes. Bacteria of the *Ideonella sakaiensis* strain [21], which involved a two-stage enzyme system in the PET degradation process, are particularly gifted in this direction. The first of them PET hydrolase (PETase) hydrolyzes PET to mono(2-hydroxyethyl)terephthalic acid (MHET), which only in the second step is hydrolyzed by MHET hydrolases (MHETases) to the starting monomers of the material, i.e. ethyl glycol and terephthalic acid. In this form, these degradation products are used as nutrients. Recently, the possibility of obtaining a PETase enzyme stable outside the cell using genetically modified *Escherichia coli* cells [22] has been shown, which could provide a technology for the degradation of PET material under controlled industrial conditions. Precise molecular analysis of the process of double-enzymatic PET decomposition has been presented very recently [23] and the obtained results indicate the enormous potential for the biodegradation of plastics based on multi-enzymatic biotechnological systems.

Reports in scientific publications concerning the potential of micro-organisms in the biodegradation of polystyrene and polyethylene are equally successful. It turns out that mealworms—the larval form of the powdery mildew *Tenebrio molitor* beetle—included polystyrene in their diet [24]. Feeding mealworm larvae only with polystyrene (styrofoam) did not cause any disturbances in the physiology of this worm, and the digestion process of polystyrene involves micro-organisms inhabiting its gastrointestinal tract. It is worth noting at this point that the larvae of powdery mildew are edible insects, rich in all the components necessary in the human diet, and it is indicated as one of the potential sources allowing the nutritional needs of humans and animals to be met [25]. In this way, huge amounts of polystyrene waste could be helpful in solving food problems on Earth. In turn, polyethylene is biodegraded by caterpillars of the *Galleria mellonella* wax moth [26]. It is not fully known what mechanism is involved in this process, but it is indicated that it may be related to the diet of these caterpillars, which feed on beeswax, which contain carbon–hydrogen bonds, analogous to polyethylene.

Micro-organisms, in their enormous variety and ease of adapting to environmental conditions, are able to obtain food and reproduce under extreme conditions [27], in which animal organisms would not have a chance to survive. These successes of micro-organisms in the struggle for survival result both from the molecular mechanisms characteristic of micro-organisms [28] and from the evolutionary adaptation to new conditions as a result of the occurring mutations [29]. It is this natural adaptation that indicates the potential use of micro-organisms in various technologies [30], including the biodegradation of hydrocarbon polymers, regardless of their structure. This thesis is supported by the fact that strains of micro-organisms capable of living in extreme conditions are usually obtained from samples taken from places characterized by these extreme conditions, including places contaminated with petroleum products [31] and plastics [21, 24]. It can be assumed that this is due to the adaptive strategy of micro-organisms taking advantage of uninhabited environmental niches and new carbon sources.

The above confirms the thesis that nature can protect us. Along with the increasing amount of environmental pollutants, micro-organisms appear that use these pollutants as a source of food. Most likely, in the near future, technologies will be available that will enable the decomposition of plastics into simple substances used by living organisms. However, the controlled use of new biodegradation technologies, in particular those based on the use of consortia of micro-organisms, should be kept in mind. It may turn out that the appetite of new micro-organisms and favorable conditions for their development may lead to uncontrolled decomposition processes of all polymer products in our environment, thus precluding the use of these useful products. It is worth noting the paper discussing the implications of the adaptation of micro-organisms to the presence of organic chemicals and the stability of these compounds [32]. Treating high-molecular polymers as organic chemicals, these remarks also apply to plastics. It seems that safer technologies would be the use of enzyme cocktails capable of specifically digesting specific chemical bonds in polymer macromolecules. In this case, there is no risk of uncontrolled reproduction of enzymes, which could occur in technological systems using living micro-organisms.

2.5 Micro-organisms in technological processes

Micro-organisms are successfully used in technological processes for the production and modification of materials, including nanomaterials. Earlier, we mentioned the use of fungi for etching PET fibers for surface modification of these fibers [20]. Fungi have also been reported as organisms capable of etching graphite and modifying diamond-like layers, reducing the graphite content and increasing the surface micro-roughness [33]. Biotechnological methods of enrichment of metal ores, even with a low concentration of these metals, offer high economic profitability of industrial processes, low energy consumption, and a significant reduction of chemical waste compared to classical methods. The fungi used for the extraction of metals and non-metals from minerals are definitely the most effective in this respect, as well as for the recovery of these materials from industrial and municipal waste, and the list of metals and metalloids obtained in this way is very extensive [34]. Equally good effects of obtaining and recovering metals are achieved with the use of bacteria on an industrial scale [35].

However, the most promising applications at present seem to be the use of micro-organisms for the production of nanoparticles. Since nanotechnologies are a very important part of material technologies and the technology of nanomaterials has been developed relatively recently, let us first discuss the most important information from the point of view of our narrative. We observe the widespread use of nanotechnology products in all areas of the economy. Unfortunately, the available physicochemical analytical techniques do not always allow for unequivocal confirmation of the presence of nanotechnology products in the offered product. There is a real threat of the presence of nanotechnology products in products, the manufacturer of which wants to hide this fact. On the other hand, there are situations when manufacturers use nanotechnology only for marketing purposes, taking advantage of the analytical difficulties in this area. Two problems overlap, namely, not entirely clear and effective legal regulations in the field of placing nanotechnology products on the market are accompanied by limited effectiveness of determining the presence of these materials in the offered products and difficulties in assessing the risk resulting from the presence of these materials.

2.6 Problems with nanomaterials

The definition of a nanoparticle (nanostructure) used for research purposes is intuitive. An object with at least one dimension in the range 1–100 nm is treated as a nanostructure. In line with this, we distinguish point nanostructures (e.g. quantum dots), one- and two-dimensional nanostructures (e.g. linear polymers, layers with nanometric thickness, nanostructured surfaces), and three-dimensional nanostructures (e.g. nanocrystals, nanotubes, dendrimers, and other spatial nanostructures). However, what is sufficient for the use of research laboratories is not always accepted by decision-makers and economic entities. Therefore, the term nanomaterial, now supplemented with the term nanoform, has been added to the terms nanoparticle and nanostructure.

The definition of nanomaterial proposed by the European Commission has the following wording (http://data.europa.eu/eli/reco/2011/696/oj):

> 2. 'Nanomaterial' means a natural, incidental or manufactured material containing particles, in an unbound state or as an aggregate or as an

agglomerate and where, for 50% or more of the particles in the number size distribution, one or more external dimensions is in the size range 1 nm–100 nm. In specific cases and where warranted by concerns for the environment, health, safety or competitiveness the number size distribution threshold of 50% may be replaced by a threshold between 1 and 50%.

3. By derogation from point 2, fullerenes, graphene flakes and single wall carbon nanotubes with one or more external dimensions below 1 nm should be considered as nanomaterials.

It is worth paying attention to the great freedom in considering substances as nanomaterials. In the first part of the definition, it is assumed that 50% or more of the particles in the numerical size distribution are to meet the nanostructure condition. However, in the next part, it is allowed to consider a substance containing a much lower concentration of nanostructures as a nanomaterial, which is explained by the competitiveness of the product, i.e. the interest of the economic entity participating in the market game. The definition of a nanoform slightly improves the situation (https://echa.europa.eu/documents/10162/23047722/appendix_nanoforms_draft_to_peg_en.pdf/bb84f0ca-7688–5293–604e-fb43982c7afd):

> According to Annex VI of the REACH Regulation, a 'nanoform' is a form of a natural or manufactured substance containing particles, in an unbound state or as an aggregate or as an agglomerate and where, for 50% or more of the particles in the number size distribution, one or more external dimensions is in the size range 1 nm–100 nm, including also by derogation fullerenes, graphene flakes and single wall carbon nanotubes with one or more external dimensions below 1 nm.

The European Union has a uniform system for the registration of chemical substances, REACH (Registration, Evaluation, Authorization and Restriction of Chemicals), introduced and supervised by the European Chemical Agency (ECHA). The REACH regulation defines the methods of identification, classification, and registration of chemical substances placed on the market, however, to date, REACH does not effectively classify nanotechnology products. Additionally, REACH does not require the registration of a substance placed on the common market in the amount of up to 1 ton per year. With regard to nanostructures, this is often a quantity that exceeds the production capacity and market needs, and substances that are nanotechnology products do not need to be registered in such a situation.

The difficulty indicated above is the problem with the unambiguous identification of nanostructures. Relatively simple nanostructures (particles of metals, metal oxides, minerals, organic pigments, etc) can be relatively easily identified and characterized by physicochemical methods, such as electron microscopy (transmission electron microscopy (TEM) and scanning electron microscopy (SEM)), atomic force microscopy (AFM) and its many variants, optical methods (dynamic light scattering (DLS), light absorption (LA)), x-ray methods (small-angle x-ray scattering (SAXS), chemical composition (elemental) analysis (EA), x-ray photoelectron spectroscopy (XPS)), mass spectrometry methods (secondary ion mass spectrometry (SINS)), fractionation (field flow fractionation (FFF), size exclusion liquid chromatography (SEC)), and free surface measurement (adsorption of inert gas (BET method)). However, even these highly specialized methods prove insufficient for complex nanostructures, in particular for multi-element composition and complex internal structures (for example synthetic biopolymers and hybrid structures). For such situations, other, alternative methods of detecting, identifying, and characterizing nanostructures should be sought, also with the use of biosensors and molecular biology methods. In the current book, in chapter 4 'Biological recognition of materials' and chapter 5 'Quantum basis of materials recognition by cells', the potential possibilities of using molecular biology methods and quantum physics techniques for these purposes are presented.

Why do nano-scale products have completely different, surprising, and often unpredictable properties compared to the same substance observed in the macroscale? One of the main reasons is the geometric factor, understood as the ratio of the surface area of a given structure to its volume. Along with the decrease in the linear size of the particles of the substance, the ratio of the surface area of these particles to their volume increases, and the increase becomes rapid when the linear dimensions of the nanostructure drop below 100 nm. This means that an increasing percentage of the atoms of the substances are exposed on the surface of the nanostructure, up to the limit conditions, when all atoms are available on the surface of the nanostructure. This situation favors significant changes in the physical properties of substances at the nano-scale and the accompanying changes in chemical reactivity, as well as the widespread occurrence of quantum effects not observed on the macroscale. As an example of the influence of a geometric factor on the observed physical processes, let us consider the cell membrane surrounding a living animal cell (figure 2.3).

The thickness of this membrane is about 10 nm, while the resting membrane potential is about 100 mV. The membrane can be considered as a phospholipid bilayer enriched with integral and peripheral proteins. The cell membrane is a structure resembling a flat capacitor, the facings of which are polar groups of phospholipids, while the dielectric layer is made of hydrocarbon chains. For the geometric and electrical conditions defined in this way, an electric field strength of about 10^7 V m^{-1} is present through the cell membrane.

This is a powerful electric field that is difficult to achieve and poses a serious threat to living organisms in the macroscopic world. On the other hand, on the nano-scale it is capable of a strong polarization of the dielectric which is inside a cell membrane, and thus stiffening this membrane. Under resting conditions the spontaneous thermal movement of hydrocarbon chains is strongly limited, and transport across the cell membrane is possible only with the use of specialized protein transport systems. The stimulation of the cell leads to the depolarization of the cell membrane and an increase in the fluidity of the phospholipid bilayer, which

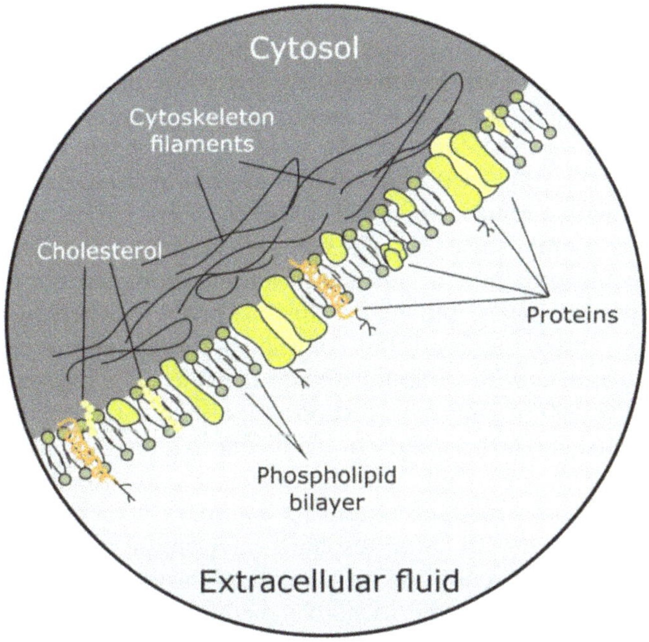

Figure 2.3. Schematic of the cell membrane.

enables the movements of the components of the cell membrane to facilitate the transport of various substrates through the membrane, also by simple diffusion.

It is also worth noting that the nano-scale is a typical scale for the size of the regions near the membrane surface, where localized biochemical reactions take place. In the cell, the reactions are located in very small volumes of attoliters (10^{-18} L), and micromolar changes in the concentration of substrates and products correspond to the entry into the reaction or flows of several dozen (about 60) molecules, while on the macroscopic scale corresponding to *in vitro* experiments in a test tube this corresponds to approximately 6×10^{17} molecules. It is obvious that it is easier to control the reactions of several dozen molecules or ions in a cell than the reactions of a billion billions of these molecules in a test tube. It is also often the cause of misunderstandings in the proper understanding of the course of physiological processes.

However, in order not to leave the impression that the nano-scale applies only to biological objects, it is worth realizing that to cover the surface of a square with a side of 100 m (10^4 m^2) with a monolayer of graphene, about 10 g of this material is needed, the total volume of which is about 5 mL (5×10^{-6} m^3). For comparison, about 2700 kg and a volume of 1000 L (1 m^3) of aluminum are needed to cover the same surface with 0.1 mm thin aluminum foil. Comparing the strength of such coatings clearly shows the advantage of graphene over aluminum as graphene has a tensile strength of about 130 GPa compared to about 350 MP for aluminum. The nano-scale geometric effects are inseparably linked with quantum effects, unobserved in the macroscopic world. One can cite a number of examples in this area, but for reasons of brevity, only those which are relatively easy to observe, sometimes even with the naked eye, are discussed, as in the case of a suspension of gold nanoparticles showing different colors in white light depending on the size of these particles and their shape [36]. For nanorods, the color of the suspension changes with the ratio of length to diameter (the nanorods' aspect ratio), for nanospheres the color changes with the thickness of the gold coating (shell thickness), and for nanocages the color changes with the content of gold in these nanostructures. Also for gold particles with a regular spherical shape, the color of the suspension changes with the size of the nanoparticle [37]. The color effect of the nanoparticles suspension is related to the phenomenon of surface plasmon resonance taking place on the surface of the nanoparticles and in the surface layer. According to the laws of geometric optics, under the conditions of total internal reflection of light at the boundary of two media with different optical densities, the energy of the incident light is reflected by 100%. However, a careful analysis of the effect shows that the incident light penetrates the boundary of the media to a depth of about 300 nm, and the photons of the incident light can be resonantly absorbed by free electrons (plasmons) located in this medium and at its boundary, which results in an energy deficit in the reflected light stream at resonant wavelength, which manifests itself in the colored effects of light scattering on the colloid particles. This deficit is so large that it can be measured with the available methods of light refractometry. The conditions of resonant photon absorption depend on the dielectric properties of an optically less dense medium. This phenomenon was used to observe mass changes at the boundary surface of both media, obtaining an extraordinary sensitivity of the measurement of one pg/cm^2 [38]. As another example, we can consider changes in the width of energy bands in objects smaller than 100 nm. These bands arise as a result of the overlapping and degeneration of the energy levels of electrons in atoms that make up the structure of a solid. Under macroscopic conditions, the number of atoms contributing to the emerging energy band is very large, while in nanostructures this number significantly decreases, which results in changes in the width of the valence, forbidden, and conductivity bands. The effects of this phenomenon are changes in the electric current conduction properties and changes in the wavelength of light adsorbed by nanostructures [39]. The above geometric and quantum effects, located on the nano-scale, were also analyzed previously by the authors in the local journal *Laboratorium—Przegląd Ogólopolski* [40].

Natural nanostructures have been with us since time immemorial. These are products created during volcanic eruptions, in the course of erosive processes or as a result of wear of materials in the process of friction. The nanostructures present in living organisms cannot be ignored either. Some of these nanostructures (nanoparticles) arise as a result of purely physical processes (friction wear products), some arise as a result of mixed physicochemical processes (erosion, volcanic eruptions), and others as a result of chemical processes (membrane structures and biomolecules of living organisms).

In analogy to the methods of creating natural nanostructures, two strategies for the production of engineering nanomaterials have been developed. The first is the grinding of the macroscopic substance until the desired particle size is obtained. This strategy is called 'top down' and uses either physical methods or a combination of physicochemical methods. The opposite strategy, called ‘bottom up’, uses chemical methods and manipulates individual atoms and groups of atoms in order to obtain the intended nanostructure. The top down methods provide homogeneous or relatively uncomplicated nanostructures at a relatively low cost. On the other hand, the bottom up chemical synthesis methods enable the production of both simple homogeneous nanoparticles, also possible to obtain by crushing methods, as well as complex nanostructures with different types of atoms arranged in the intended manner in this nanostructure, which are not available using the top down methods.

Traditionally and widely used technologies for the production of metallic nanoparticles are based on chemical synthesis in the liquid phase with the use of various reagents, in particular reducing substances, such as hydrazine [41], potassium bitartrate [42], or sodium borohydride [43]. The stability of the resulting nanoparticles in suspension is achieved by the use of various stabilizing substances, such as polyvinylpyrrolidone [43] or, more and more often, citric acid [44]. Polymer nanoparticles with a more complex structure and different chemical composition are produced using a dispersion of preformed polymers and the polymerization of monomers [45]. Chemical methods of solid-phase organic synthesis of organic peptide polymers, nucleic acids, and complex sugars are also used [46]. While these methods are economically viable, their use generates significant amounts of toxic reagents, solvents, and intermediates. Therefore, currently technologies for the synthesis of nanoparticles using biotechnology are increasingly being sought. These methods are available for the synthesis of metallic nanoparticles (gold, silver, cadmium, palladium, platinum), semi-metallic (bismuth and tellurium), non-metals (selenium), metal alloys (gold-silver), and nanoparticles formed from oxides (cerium, silicon, titanium, iron, zircon, antimony, copper, and zinc) and from sulfides (cadmium, lead, iron, copper, silver, zinc, and antimony) [47]. The list of possible nanoparticles synthesized by biotechnological methods is not limited to those mentioned above and the reader will surely find more examples. Taking into account the ability of living organisms to synthesize much more complex polymeric organic nanoparticles, it can be hoped that biotechnology will allow the production of any complex organic polymer nanoparticles with the assumed physicochemical properties and biological activity in the near future.

2.7 Threats from nanotechnology products

The use of clean technologies for the production of nanomaterials, based on the use of biotechnology, significantly reduces the environmental burden, potentially allows the use of these technologies without restrictions due to the level of industrialization of the region or country, which may be a factor that equalizes the development opportunities of poorer regions, but does not affect the threat resulting from the presence of nanotechnology products in our environment. What is the undoubted technological advantage of nanomaterials (nanoparticles or nanoforms) becomes a serious threat to living organisms. Such small objects can easily diffuse in the environment and penetrate through biological barriers such as cell membranes and the larger barrier structures of multicellular organisms that separate the organism from the environment. These include the epithelium of the oral cavity and respiratory tract, lungs, stomach, intestines, and the endothelium of blood vessel walls in close co-operation with the epithelium. After all, nutrients and regulatory substances, as well as products of cellular metabolism, penetrate this way. All macromolecules involved in the life processes of organisms are sized in the nanometric range and their highly specialized physiological functions are associated with their small size. One cannot exclude in advance the possibility of accidental production of engineering nanostructures showing analogous biological activity, or worse, an activity that is not manifested by natural biomolecules, and the appearance of which, as a result of contact of a biological object with a nanostructure, may result in previously unknown effects. It cannot be ruled out that this may lead to the emergence of new diseases of unknown and as yet unidentified etiology. The possibility of producing nanoparticles with chemically clean biotechnological methods does not mean that the manufactured nanomaterials are completely safe. The risk resulting from exposure of living organisms to contact with nanomaterials is still not fully defined, but there is no doubt that such a risk exists [48].

Our current incomplete knowledge of the nanostructures themselves and their interactions with biological objects is somewhat analogous to the incomplete knowledge of the phenomenon of radioactivity that researchers had at the initial stage of the development of knowledge about this phenomenon. Both the products of radioactivity and the products of nanotechnology elude our senses. We do not see them, we do not feel them, we do not recognize them by smell or taste, but these products pose a serious threat to living organisms. In relation to the phenomenon of radioactivity, the physical laws applicable to this day were formulated relatively quickly and radiation detectors were constructed that recognize both the types of radiation and the energy carried by radiation. The radioactivity researchers had a slightly easier task compared to current nanomaterial researchers. First of all, in the phenomena of radioactivity we deal with only a few types of radiation, excluding of course the short-lived elementary particles that, as a result of conversions and decays, lead to the formation of these basic types of radiation. We mean corpuscular alpha, beta, proton, and neutron radiation as well as electromagnetic and neutrino radiation, with all the complexity of interaction of this radiation with matter. However, the knowledge of the ways in which radiation interacts with matter, i.e. with atoms and their components, allows for relatively simple predictions of the effects of the interaction of this radiation with living organisms, which are, after all, made of the same atoms. It allows to predict the destruction of biopolymer structures and changes in their biological functions, which are clearly confirmed in experimental studies. This knowledge also allowed the construction of radiation detectors allowing for quick warning against invisible danger, and in the event of exposure of the body to radiation, to determine the type of radiation and the absorbed dose, which enables the implementation of appropriate medical procedures. As in the case of natural nanoparticles, living organisms have been exposed to radiation from the beginning of their existence and this radiation has had and still has a significant impact on the appearance of mutations in the genetic material of cells and thus on the course of evolution of living organisms.

At present, it is difficult to predict what effect the exposure of organisms to nanotechnology products may have on living organisms. The situation is complicated to a high degree by the virtually unlimited variety of nanoparticles and nanostructures resulting from differences in atomic composition, as well as the different sizes and spatial structure of these nanotechnology products, as well as an acute shortage of comprehensive knowledge regarding the interaction of these products with biological objects. So far, it has not been possible to develop an effective method of grouping nanomaterials due to their characteristic features and linking these features with the nature of the biological effects induced, although such attempts are still being made using nanoQSAR models [49] and read-across strategies [50]. Unfortunately, the analogy with the analysis of the interaction of radiation with matter does not apply in this case.

Nevertheless, there are attempts to use living organisms as indicators of the presence of nanotechnology products. Worth noting are model multicellular organisms, in particular the nematode *Caenorhabditis elegans* (*C. elegans*), which was successfully used to assess the *in situ* absorption and toxicity of silver nanoparticles in a specially constructed microfluidic chip [51], which makes it possible to produce a biological indicator of the presence of silver nanoparticles. The authors of another study [52] were heading in a similar direction, demonstrating that *C. elegans* is an organism very sensitive to environmental changes and able to respond in a specific way to exposure to some nanomaterials, not only to nanosilver, which allows considering this organism to be used as a biosensor for early warning against the presence of certain nanoparticles. Attempts are being made to use other organisms to observe the effects of their exposure to nanomaterials using an approach of high content analysis for a large amount of experimental data [53]. It is hoped that these efforts, taking into account the latest advances in molecular research technologies, will bring the expected results and it will be possible to both track the presence of nanostructures and predict the effects of contact of living organisms with these products.

It is also worth recalling the experience of human society resulting from the history of discovery and subsequent use as well as the ecological effects of the use of dichlorodiphenyltrichloroethane (DDT). DDT is an organic chemical compound belonging to the group of chlorinated hydrocarbons, first synthesized in 1874 by Othmar Zeidler (1850–1911) and rediscovered by Paul Müller (1899–1965), and used for its strong insecticidal properties. DDT was widely used by the Allied Forces during World War II to combat typhus and malaria carrier insects. It is estimated that the use of DDT contributed to a significant reduction in disease among soldiers, which obviously translated into a reduction in human casualties and had a significant impact on the outcome of hostilities.

In recognition of his services in this regard, Paul Müller was awarded the Nobel Prize in 1948. This is one very positive face of DDT. Unfortunately, there is a second, very negative aspect of this compound widely used as a plant protection product, which has contributed to very dramatic ecological changes. This product is very persistent, with a half-life of several years, and its degradation products are even more persistent and equally toxic to living organisms. Initially, the extinction of fish and birds was observed in areas of intensive use of this substance, and later, scientifically justified toxic effects on higher organisms, including humans, were described. It was not until the 1960s that the use of DDT was legally limited, but the negative effects of its earlier use are still observed. Unfortunately, DDT is no exception, many other persistent chemicals are found in significant amounts in our environment and accumulate in our bodies. In this case, it seems that nature is helping us, because these chlorinated chemical compounds turn out to be effectively neutralized and removed from the environment by micro-organisms [54].

New material technologies, based on the achievements of materials science, have made technological progress possible and have contributed significantly to the evolution and breakthroughs in social, industrial, and economic development. It was the new materials and knowledge about their production and their properties that made it possible to build the first steam engine, build electric machines, and electrify industry and individual households, automate production, as well as invent and mass-produce semiconductor components, without which the development of information technology and the emergence of the Internet would have been impossible. Without the achievements in the field of new materials technologies, progress in the field of life sciences would not be possible and new technologies for the bioeconomy, medical diagnostics, and for the currently commonly used clinical technologies would not be possible. On the other hand, it is the new material technologies and the economic development they generate that are largely responsible for the current ecological and climate problems. It seems that only close co-operation of the representatives of materials science with representatives of life sciences, with the active support of information technologies, can lead to creation of new technologies for the production of modern materials and new technologies aimed at reusing industrial and municipal waste as part of waste recycling and the cyclic economy. It is life sciences that currently provide development impulses for new technologies for the production and processing of industrial products and for monitoring the state of the natural environment and for the safe use of modern technology products. Thus, it can be concluded that the era of biology in materials science has already arrived and, together with physics and chemistry, is the fundamental triad of new material technologies.

References

[1] Andres C *et al* 2017 The Arctic Ocean as a dead end for floating plastics in the North Atlantic branch of the thermohaline circulation *Sci. Adv.* **3** e1600582

[2] Claire M *et al* 2020 Thirty years of marine debris in the Southern Ocean: annual surveys of two island shores in the Scotia Sea *Environ. Int.* **136** 105460

[3] Horton A A, Walton A, Spurgeon D J, Lahive E and Svendsen C 2017 Microplastics in freshwater and terrestrial environments: evaluating the current understanding to identify the knowledge gaps and future research priorities *Sci. Total Environ.* **586** 127–41

[4] Carr S A, Liu J and Tesoro A G 2016 Transport and fate of microplastics in wastewater treatment plants. *Water Res.* **91** 174–82

[5] Eerkes-Medrano D, Thompson R C and Aldridge D C 2015 Microplastics in freshwater systems: a review of the emerging threats, identification of knowledge gaps and prioritisation of research needs *Water Res.* **75** 63–82

[6] European Commission, *6 Commission priorities for 2019–24* https://ec.europa.eu/info/strategy/priorities-2019-2024_en

[7] Walkowiak B, Walkowiak-Przybylo M and Komorowski P 2019 Materiomika—pomost łączący nauki o materiałach z naukami życiu (Materiomics—a bridge connecting materials science with life sciences) *Laboratorium—Przegląd Ogólnopolski* **2** 24–30 (in Polish)

[8] Xu M, David J M and Kim S H 2018 The fourth industrial revolution: opportunities and challenges *Int. J. Finan. Res.* **9** 90–5

[9] Statistics Times *Gross World Product* http://statisticstimes.com/economy/gross-world-product.php

[10] Salomons W and Forstner U (ed) 1988 *Chemistry and Biology of Solid Waste* (Berlin: Springer)

[11] Yohannes H and Elias E 2017 Contamination of rivers and water reservoirs in and around Addis Ababa city and actions to combat it *Environ. Pollut. Climate Change* **1** 116

[12] Coe H 2020 Airborne particles might grow fast in cities *Nature* **581** 14

[13] Miejskie Przedsiębiorstwo Wodociągów i Kanalizacji w Warszawie (Municipal Water and Sewerage Company in Warsaw) *Małże i ryby testerami wody (Mussels and fish are water testers)* https://mpwik.com.pl/view/malze-i-ryby-testerami-wody&i=678

[14] Chen J, Gu W, Tao J, Xu Y, Wang Y, Jingyan G and Du S 2017 The effects of organic residue quality on growth and reproduction of *Aporrectodea trapezoides* under different moisture conditions in a salt-affected agricultural soil *Biol. Fertil. Soils* **53** 103–13

[15] Martinez-Salgado M M, Gutiérrez-Romero V, Jannsens M and Ortega-Blu R 2010 Biological soil quality indicators: a review *Current Research, Technology and Education Topics in Applied Microbiology and Microbial Biotechnology* ed A Mendez-Vilas (Badajoz: Formatex) pp 319–28

[16] Sumampouw O J and Risjani Y 2014 Bacteria as indicators of environmental pollution: review *Int. J. Ecosyst.* **4** 251–8

[17] Szczepaniak K and Biziuk M 2003 Aspects of the biomonitoring studies using mosses and lichens as indicators of metal pollution *Environ. Res.* **93** 221–30

[18] Nowak D J, Hirabayashi S, Bodine A and Greenfield E 2014 Tree and forest effects on air quality and human health in the United States *Environ. Pollut.* **193** 119–29

[19] Sanderfoot O V and Holloway T 2017 Air pollution impacts on avian species via inhalation exposure and associated outcomes *Environ. Res. Lett.* **12** 083002

[20] Nimchua T, Eveleigh D E, Sangwatanaroj U and Punnapayak H 2008 Screening of tropical fungi producing polyethylene terephthalate-hydrolyzing enzyme for fabric modification *J. Ind. Microbiol. Biotechnol.* **35** 843–50

[21] Tanasupawat S, Takehana T, Yoshida S, Hiraga K and Oda K 2016 *Ideonella sakaiensis* sp. nov., isolated from a microbial consortium that degrades poly(ethyleneterephthalate) *Int. J. Syst. Evol. Microbiol.* **66** 2813–8

[22] Seo H, Kim S, Son H F, Sagong H-Y, Joo S and Kim K-J 2019 Production of extracellular PETase from *Ideonella sakaiensis* using sec-dependent signal peptides in *E. coli Biochem. Biophys. Res. Commun.* **508** 250e255

[23] Knotta B C *et al* 2020 Characterization and engineering of a two-enzyme system for plastics depolymerization *Proc. Natl Aacd. Sci.* **117** 25476–85

[24] Yang Y, Yang J, Wu W-M, Zhao J, Song Y, Gao L, Yang R and Jiang L 2015 Biodegradation and mineralization of polystyrene by plastic-eating mealworms: part 1. Chemical and physical characterization and isotopic tests *Environ. Sci. Technol.* **49** 12080–6
[25] van Huis A, Itterbeeck J V, Klunder H, Mertens E, Halloran A, Muir G and Vantomme P 2013 Edible insects: future prospects for food and feed security *FAO Forestry Paper* 171 Food and Agriculture Organization of the United Nations, Rome
[26] Bombelli P, Howe C J and Bertocchini F 2017 Polyethylene biodegradation by caterpillars of the wax moth *Galleria mellonella Curr. Biol.* **27** R292–3
[27] Arora N K and Panosyan H 2019 Extremophiles: applications and roles in environmental sustainability *Environ. Sustain.* **2** 217–8
[28] Bleuven C and Landry C R 2016 Molecular and cellular bases of adaptation to a changing environment in microorganisms *Proc. R. Soc.* B **283** 1458
[29] Sprouffske K, Aguilar-Rodríguez J, Sniegowski P and Wagner A 2018 High mutation rates limit evolutionary adaptation in *Escherichia coli PLoS Genet.* **14** e1007324
[30] Lenski R E 2017 What is adaptation by natural selection? Perspectives of an experimental microbiologist *PLoS Genet.* **13** e1006668
[31] Adebusoye S A, Ilori M O, Amund O O, Teniola O D and Olatope S O 2007 Microbial degradation of petroleum hydrocarbons in a polluted tropical stream *World J. Microbiol. Biotechnol.* **23** 1149–59
[32] Poursat B A J, van Spanning R J M, de Voogt P and Parsons J R 2019 Implications of microbial adaptation for the assessment of environmental persistence of chemicals *Crit. Rev. Environ. Sci. Technol.* **49** 2220–55
[33] Kaczorowska A, Szczęsna-Antczak M, Antczak T, Kaczorowski W, Niedzielski P, Mitura S and Kozanecki M 2006 An influence of microbial viable cells on diamond-like carbon films *NANODIAM—New Technologies for Medical Application: Studying and Production of Carbon Surfaces Allowing for Controllable Bioactivity* ed S Mitura, P Niedzielski and B Walkowiak (Warsaw: Państwowe Wydawnictwo Naukowe) pp 99–116
[34] Liang X and Gadd G M 2017 Metal and metalloid biorecovery using fungi *Microb. Biotechnol.* **10** 1199–205
[35] Devasia P and Natarajan K A 2004 Bacterial leaching: biotechnology in the mining industry *Resonance—J. Sci. Edu.* **9** 27–34
[36] Dreaden E C, Alkilany A M, Huang X, Murphy C J and El-Sayed M A 2012 The golden age: gold nanoparticles for biomedicine *Chem. Soc. Rev.* **41** 2740–79
[37] Link S and El-Sayed M A 1999 Size and temperature dependence of the plasmon absorption of colloidal gold nanoparticles *J. Phys. Chem.* B **103** 4212–17
[38] Biacore 2003 *Biacore® Sensor Surface Handbook* BR-1005-71 GE Healthcare p 31
[39] Irshad Ahamed M and Sathish Kumar K 2019 Modelling of electronic and optical properties of Cu_2SnS_3 quantum dots for optoelectronics applications *Mater. Sci.-Poland* **37** 108–15
[40] Walkowiak B 2014 Nanotechnologia—korzyści i zagrożenia (Nanotechnology—the benefits and risks) *Laboratorium* **3–4** 58–63
[41] Li Y, Duan X, Qian Y, Yang L and Liao H 1999 Nanocrystalline silver particles: synthesis, agglomeration, and sputtering induced by electron beam *J. Colloid Interface Sci.* **209** 347–9
[42] Tan Y, Dai X, Li Y and Zhua D 2003 Preparation of gold, platinum, palladium and silver nanoparticles by the reduction of their salts with a weak reductant–potassium bitartrate *J. Mater. Chem.* **13** 1069–75

[43] Kim J S *et al* 2007 Antimicrobial effects of silver nanoparticles *Nanomed.: NBM* **3** 95–101
[44] Shinohara S, Eom N, Teh E-J, Tamada K, Parsons D and Craig V S J 2018 The role of citric acid in the stabilization of nanoparticles and colloidal particles in the environment: measurement of surface forces between hafnium oxide surfaces in the presence of citric acid *Langmuir* **34** 2595–605
[45] Rao J P and Geckeler K E 2011 Polymer nanoparticles: preparation techniques and size-control parameters *Prog. Polym. Sci.* **36** 887–913
[46] Parsons J G, Sheehan C S, Wu Z, James I W and Bray A M 2003 A review of solid-phase organic synthesis on SynPhase™ lanterns and SynPhase™ crowns *Meth. Enzymol.* **369** 39–74
[47] Tiquia-Arashiro S and Frigi Rodrigues D 2016 *Extremophiles: Applications in Nanotechnology* (Berlin: Springer)
[48] Klaine S J, Koelmans A A, Horne N, Carley S, Handy R D, Kapustka L, Nowack B and von der Kammer F 2012 Paradigms to assess the environmental impact of manufactured nanomaterials *Environ. Toxicol. Chem.* **31** 3–14
[49] Gajewicz A, Schaeublin N, Rasulev B, Hussain S, Leszczynska D, Puzyn T and Leszczynski J 2014 Towards understanding mechanisms governing cytotoxicity of metal oxides nanoparticles: hints from nano-QSAR studies *Nanotoxicology* **9** 313–25
[50] Stone V *et al* 2020 A framework for grouping and read-across of nanomaterials supporting innovation and risk assessment *Nano Today* **35** 100941
[51] Kim J H, Lee S H, Cha Y J, Hong S J, Chung S K, Park T H and Choi S S 2017 *C. elegans*-on-a-chip for *in situ* and *in vivo* Ag nanoparticles' uptake and toxicity assay *Sci. Rep.* **7** 40225
[52] Walczynska M, Jakubowski W, Wasiak T, Kadziola K, Bartoszek N, Kotarba S, Siatkowska M, Komorowski P and Walkowiak B 2018 Toxicity of silver nanoparticles, multiwalled carbon nanotubes, and dendrimers assessed with multicellular organism *Caenorhabditis elegans Toxicol. Mech. Methods* **28** 432–9
[53] Mittal D and Kaul G 2018 Towards nano-risk assessment with high throughput screening and high content analysis: an intelligent testing strategy *Comprehensive Nanoscience and Nanotechnology* 2nd edn (Amsterdam: Elsevier)
[54] Bhatt P, Kumar M S, Mudliar S and Chakrabarti T 2007 Biodegradation of chlorinated compounds—a review *Crit. Rev. Environ. Sci. Technol.* **37** 165–98

IOP Publishing

Biological Evaluation of Materials
The interaction of materials with their environment
Bogdan Walkowiak, Magdalena Walkowiak-Przybyło and Piotr Komorowski

Chapter 3

Materiomics

Progress in materials science and life sciences would not be possible without technical and technological innovations. This chapter discusses the importance of microscopic, spectroscopic, and other research techniques for the advancement of science. Mutual relations between materials science and the life sciences have found their place in materiomics by using the achievements of both fields of science.

3.1 The importance of microscopic techniques for the development of science

Microscopic techniques have played a role that cannot be overestimated in the advancement of knowledge, both in the field of life sciences and in the field of materials science. The invention of the microscope, around 1590, is credited to the Dutchman Zachariasz Janssen (1580–1632). However, due to the young age of Zacharias, who was a teenager at the time, a significant share in the invention is attributed to his father Hans, a respected eyewear manufacturer at the time. Some sources point to the independent and simultaneous invention of the microscope by another Dutch optician producing glasses, Hans Lipperhey (1570–1619), who was also the inventor of the optical telescope for astronomical observations. The invention of the microscope, however, was not appreciated by contemporaries. Only the improvement of the microscope construction by Antoni van Leeuwenhoek (1632–1723) and his passion for the observation of various objects, including micro-organisms and living animal and plant cells, as well as protozoa, contributed to the rapid development of biology, and Leeuwenhoek himself deserves to be called the father of microbiology. In 1683 Leeuwenhoek sent a letter to the Royal Society in London, in which he described his first microscopic observations of biological objects. An extract of this letter is available from the Royal Society [1].

Thanks to these observations and the adoption of the light microscope, researchers have obtained a powerful tool that allows for the quantitative and qualitative analysis of living organisms as well as penetration into their interior. Thanks to this, new fields of science were created—microbiology and cytology. The optical microscope also contributed to enormous advances in the understanding and treatment of dangerous infectious diseases. In the late nineteenth century, infectious diseases such as tuberculosis, rabies, plague, and cholera contributed to mass deaths. Louis Pasteur (1822–1895) and Robert Koch (1843–1910) made groundbreaking discoveries. Pasteur developed a vaccine against rabies and anthrax, as well as one of the first vaccines against cholera. Koch, in turn, discovered and described the bacterium that causes tuberculosis and worked to identify the causes of other diseases which threaten humans, including leprosy, bubonic plague, and malaria. The microscope was also used to observe genetic material in the nucleus of the cell, including the study of chromosome division in the process of cell division. In 1919 Thomas Hunt Morgan (1866–1945) proved that chromosomes are the carriers of genes, thus giving rise to genetics and starting an extremely intense period of development in the life sciences, which has continued in varying degrees to the present day.

Let us return to the science of materials. When going back in history, particularly to antiquity, one can observe a clear preference for spirit over matter, and those who, because of their profession, dealt with materials were usually treated with a great disdain, not to say with disgust. Among the beautiful and perfect gods of Greek mythology [2] the only crippled and ugly one was Hephaestus—the god of fire and patron of blacksmiths, goldsmiths, and other craftsmen who process materials. However, Hephaestus was able to combine his disability and ugliness with the beauty of his wife Aphrodite—the goddess of love—as well as with the first women he created. This mythically justifies the complexity of materials science, including the combination of beauty and utility, theory and practice, and the microworld of atoms with the macroworld of construction. With the passage of time, the advantage of spirit over matter decreased significantly and there was a permanent separation of these two concepts, which were fundamental at the dawn of history.

The history of the science of materials, their properties, and applications is as old as the history of mankind. Various breakthrough moments can be identified in this history, but one of the most important, just like in life sciences, was the invention of the microscope and its effective use. In 1858 Henry Clifton Sorby (1826–1908) published the work 'On the microscopical, structure of crystals, indicating the origin of minerals and rocks' [3], in which he described microscopic observations of natural and artificial crystals and indicated characteristic structures that link them with the conditions of crystallization.

Another Sorby observation from 1863, recorded in a laboratory journal, concerned the observation of the structure of steel. The use of a microscope for systematic studies of the structure of steel was described by Sorby in 1886 in the work 'On the application of very high powers to the study of the microscopical structure of steel' [4]. Since then, the microscope has become a widely used research device in materials science. Of course, the success of using the microscope in materials research was largely determined by the ability to use it, resulting from the knowledge of the structure of this device and its operation, as well as the ability to prepare samples of materials for analysis.

The following years and subsequent research challenges contributed to the development of microscopic techniques. In order to remove the optical defects of simple lenses (spherical and chromatic aberrations), complex lens systems were introduced to correct these defects. Additionally, the introduction of strong illumination of the tested sample allowed to obtain a sharp, contrasting image of the observed object both in reflected light and in the light passing through the observed object. A number of modifications were introduced, including the possibility of phase-contrast observations, observations in a dark field, the use of fluorescence phenomena, as well as obtaining confocal images (focused in selected planes). These modifications widened the possibilities of applying optical microscopy enormously in both life sciences and materials science. Unfortunately, these modifications did not allow the limited resolving power of the light microscope to be overcome, which depends on the wavelength of the light in which the observation is made, and this in turn limits the possibilities of magnifying the image obtained with the use of the microscope. The best optical microscopes, operating with the use of polarized near ultraviolet light, achieve a maximum magnification of about 3500×. Visible light microscopes achieve maximum magnifications of about 1500×.

The natural idea of using electromagnetic radiation with a shorter wavelength to improve the microscope resolution and obtain higher magnifications then appeared. X-ray radiation, despite being an electromagnetic wave just like light radiation, is not visible to the human eye and interacts with matter in a different way, due to the much higher energy of the photons. The lack of eye response can be replaced by using an x-ray film or a charge-coupled device (CCD) detector to create an image. On the other hand, even soft x-rays are absorbed much more than reflected by the materials they fall on, which makes it difficult to make an optical system (ocular and objective) to form a beam of this radiation. Despite these difficulties, soft x-ray microscopy systems have been developed [5], although they are not widely used.

A slightly different strategy was applied to the construction of the optical ultramicroscope. The illumination of the examined object by strong light incident perpendicular to the optical axis of the microscope was used [6]. The device, invented in 1903, uses the Tyndal effect (light scattering by a colloid to form a characteristic light cone) to study colloidal particles much smaller than the wavelength of visible light. This approach did not allow for the determination of the shape of the suspended particles, but it allowed to determine the number of these particles per unit of volume, which allowed the calculation of the size of these particles. The invention of the optical ultramicroscope contributed to the development of colloid chemistry and significantly contributed to the possibility of examining particle suspensions with a size of single nanometers, so it can be safely stated that it was the first device for studying nanoparticles and contributed to the development of nanotechnology.

Work is ongoing to improve optical microscopy, in particular for biomedical applications. An interesting solution that allows more than twice the resolving power of a classic light microscope is scanning super-resolution microscopy, based on the measurement of the correlation of fluctuations of fluorescence brightness as image contrast [7].

The next step in the progress of imaging objects was the invention of the electron microscope [8]. According to the de Broglie hypothesis, a moving electron can be assigned a wave of matter whose length depends on the energy of the electron's motion. The matter wave associated with the movement of an electron is subject to the phenomena of reflection, diffraction, interference, and absorption by the material object on which it falls, just like an electromagnetic wave in the visible range in a light microscope. The faster the electron, the shorter the wavelength, and this also means the reduction of the resolution limit as a function of the electron's energy, which in turn translates into the possibility of obtaining many times greater magnifications of the observed object. In 1931 Ernst Ruska (1906–1988) constructed the first transmission electron microscope (TEM). The idea of the structure and operation of this microscope is very similar to an optical microscope used to observe objects in the light passing through the tested sample, with the difference that instead of a light beam, a beam of electrons accelerated to the appropriate energy is used, and instead of glass lenses, the focusing of the electron beam is achieved with the help of magnetic field (magnetic lenses), while the image of the observed sample was created on an x-ray film or other photosensitive devices. A few years later, in 1935, Max Knoll (1897–1969) built the first scanning electron microscope (SEM) whose principle of operation was analogous to that of a light microscope operating in the reflected light mode. The SEM uses similar solutions for generating, accelerating, and forming the electron beam as in the TEM. In both types of microscopes, it was necessary to maintain a relatively high vacuum so that the electron beam would not be scattered on air molecules, which significantly increased the operating costs of these devices compared to optical microscopes. In addition, due to the high resolution and the magnification achieved, even small vibrations of the microscope structure translate into a deterioration of the quality of the obtained image. In order to eliminate these inconveniences, appropriate building structures are necessary to insulate the microscope system from mechanical vibrations from the environment. As in light microscopy, also in the case of electron microscopy, the success of imaging is determined by the quality of the microscope, but to a much greater extent by the quality of the samples prepared for observation. In the case of TEM, the thickness of the imaged sample should not exceed one micrometer, while in the case of SEM, non-conductive samples need to be sputtered with a very thin layer of conductive material in order to avoid electrostatic charging of the sample and resulting image distortion.

The TEM entered research laboratories almost immediately after its construction. The SEM, on the other hand, became commercially available only in the 1970s. Both imaging techniques using an electron beam have found a wide range of applications both in materials science [8–10] and in the life sciences [11–14]. In both areas, ultrafast electron microscopy has been used to study very fast transient processes [15]. However, also in the case of electron microscopy, the wavelength of the accelerated electron is the limit of the microscope resolution and the related magnification of the observed object. High-resolution electron microscopy allows one to distinguish and observe objects no smaller than 0.5 nm and magnification up to 1–2 million times in the case of SEM and 0.05 nm and magnification of 50 million times in the case of TEM. SEM allows one to obtain a pseudo-3D image of the sample surface and analysis of the chemical composition of this surface, while TEM offers high-resolution 2D images of the internal structure of the sample and also the analysis of the spatial distribution of individual elements in the sample volume [16].

In the second half of the last century, the first scanning probe microscopes (SPMs) for surface imaging were constructed. A comprehensive discussion of the theory, practical solutions, exemplary results, and possible SPM artefacts can be found in the book *Scanning Probe Microscopy—Atomic Force Microscopy and Scanning Tunneling Microscopy* [17]. The first such device, the scanning tunneling microscope (STM), was constructed by Gerd Binnig (1947–) and Heinrich Rohrer (1933–2013) [18] for the purposes of their study of the surface of very thin oxide layers [19]. The desired spatial resolution had to reach the nanometer fraction scale. A probe (needle) was placed over the surface of a conductive sample, or the sample was sputtered with metal atoms, which could be moved in a controlled manner. The arm holding the probe was attached to the piezoelectric scanner with which control of the movement of the probe over the sample was possible.

Scanning of successive points and lines of the sample image is carried out according to the algorithm used. Usually the probe moves over the sample at a constant height and the apparatus records the value of the tunneling current—this is called constant height mode (CHM). More advanced devices work in the constant current mode (CCM) or in the constant gap width mode (CGM) using a negative feedback system. The most advanced STM systems work in the scanning tunneling spectroscopy (STS) mode, which allows to determine the dependence of the tunneling current intensity on the applied voltage. Such a procedure makes it possible to determine the electron density on the surface of the tested substance, i.e. to visualize electron orbitals [20].

The success of the STM, while being limited to the possibility of examining conductive surfaces, inspired the search for the possibility of imaging the surface of non-conductive materials using the interaction of the probe tip atoms with the sample surface atoms. In 1986 Gerd Binning submitted an invention to the Patent Office in the USA titled: 'Atomic force microscope and method for imaging surfaces with atomic resolution' [21]. This new solution in the field of the SPM technique met with even greater recognition and a number of adaptations of this technique appeared in a short time, using various variants of interactions, also taking into account the lateral forces twisting the probe lever—lateral force microscopy (LFM)—thus enabling the study of friction on the atomic scale [22], as well as atomic force spectroscopy (AFS) useful in studying the mechanical properties of a material on the atomic scale [23, 24]. SPM has found wide application in both materials science [25] and life science [26], where the authors of this study also contribute [27, 28].

3.2 Other research techniques allowing the development of science

Although microscopic techniques played a huge role in the development of science, other research techniques that use various physical phenomena allowing for qualitative and quantitative analysis of the studied objects cannot be underestimated.

Spectroscopic methods that use the analysis of the interaction of external fields with the tested sample deserve special attention. Modern science has very different spectroscopic techniques depending on the type of external fields used. The theoretical basis for signal analysis in spectroscopy, regardless of its type, is described in [29]. Due to the multitude of solutions we certainly will not list all the methods here. The most popular are methods using electromagnetic radiation, often also referred to as optical spectroscopy, although the range of radiation used also includes infrared and ultraviolet. A brief outline of the history of optical spectrometry can be found in [30]. The physical basis of this type of spectroscopy results from the interaction of electromagnetic radiation with the electrons of the sample atoms. Due to the way the spectrum is generated, atomic adsorption spectrometry (AAS) is distinguished by the incident radiation being absorbed specifically by the excited atoms and we observe the result in the form of an absorption spectrum. An example is the spectrum of sunlight after passing through the Earth's atmosphere, called the Fraunhofer spectrum. The second type is atomic emission spectroscopy (AES), when as a result of the energy being supplied, atoms are excited and during the return to the base state they emit radiation characteristic for a given atom, known as the emission spectrum. The spectra from the analysis of atomized samples are linear (discrete), while the spectra from liquids and solids are continuous. This type of spectroscopy provides very valuable information about the chemical composition of the sample, but it does not give sufficient information about the structure of the sample, and even less about the chemical bonds that exist there, although infrared and ultraviolet absorption spectroscopy provide some information about the presence of characteristic chemical groups and bonds between atoms [31]. The complement is Raman spectrometry, which uses the phenomenon of inelastic scattering of photons on the molecular structure and allows the analysis of the molecular spectra of the tested compounds with much higher resolution [31].

The group of spectroscopic techniques using the electromagnetic field also includes nuclear magnetic resonance (NMR) [32] and electron paramagnetic resonance (EPR) [33]. Both techniques are based on the resonance absorption of radio frequency electromagnetic field energy by spin-unpaired nucleons (NMR) or spin-unpaired electrons (EPR) of atoms in a strong external magnetic field.

Spectroscopic methods also include mass spectrometry (MS) [34]. In contrast to the previously mentioned spectroscopic methods that use the interaction of the electromagnetic field with the structure of the tested material, in the case of mass spectrometry the analysis of the movement of ionized particles of the tested substance is used to identify the substance and determination of the mass to charge ratio (m/z) for each molecule. The technique is based on two independent processes. In the first process, the test substance must be ionized and various techniques are used for this purpose, most commonly chemical ionization (CI), thermospray and ionization (TSI), field desorption and ionization (FDI), laser desorption and ionization in the presence of a matrix (matrix assisted laser desorption (MALDI)), electrospray ionization (ESI) and a number of other less common methods. Additional ion fragmentation is often used to improve the resolution and efficiency of the analysis. In the second process, after giving the ions kinetic energy, it is necessary to separate them according to the *m/z* value using various types of analyzers, including time of flight (TOF), ionic trap (IT), quadrupole analyzers (QA), magnetic and electric separators (MES). The separated ions still have to be registered by a detector, most often an electron multiplier or a photomultiplier. The obtained spectrum of the registered signal intensity as a function of *m/z* is then used to identify the molecules. This identification can be made on the basis of theoretical analysis and determination of the expected spectrum based on the knowledge of the chemical and isotopic composition of the molecule. However, for complex molecules such a procedure is very laborious and currently, in the vast majority of cases, identification is made using mass spectral libraries collected in commercial databases.

It is also worth mentioning the chromatographic methods of separation of a mixture of molecules [35]. A common feature of all chromatographic methods is the use of interactions between the stationary phase and the mobile phase to separate molecules. Based on the type of stationary phase and the way it is used, chromatographic methods are divided into paper chromatography, thin layer chromatography, and column chromatography, while based on the type of mobile phase we have liquid and gas chromatography. In turn, based on the physical phenomena used for the separation of molecules, chromatography can be classified into a number of techniques. Historically, the earliest introduced was the separation of molecules due to their size and mass. This type of chromatography is known as gel filtration chromatography (GFC) or size exclusion chromatography (SEC). Next, different types of adsorption chromatography were introduced. Based on the interaction of electric charges ion exchange chromatography (IEC) was developed, and based on hydrophobic properties, hydrophobic interaction chromatography (HIC) and the related reverse phase chromatography (RPC) were developed. The most selective type is affinity chromatography (AC). Examples of the application of liquid chromatography methods are described in [36].

The group of methods for the analytical or preparative separation of molecules also includes electrophoretic techniques [37]. The common feature of chromatographic methods is the presence of a stationary phase. However, unlike chromatographic methods, there is no mobile phase here, and the separation of molecules takes place thanks to the movement of electrically charged particles in the electric field. Depending on the conditions of the electrophoretic separation, different types of electrophoresis are used.

Zone electrophoresis (ZE) is the basic and most widely used type of electrophoresis. It takes place in a medium in which the electrolyte has a constant pH value throughout its entire volume. The difference in the migration distances of individual macro-ions over a specified period of time results directly from the difference in the electrophoretic mobility of these ions.

Isotachophoresis (ITP) is a type of electrophoresis in which there is discontinuity in the pH value. The macro-ions of the separated sample migrate in the area between two electrolyte systems with different pH values and different mobilities of electrolyte ions. The result of this type of electrophoresis is the ordering of the separated macro-ions according to their decreasing electrophoretic mobility. The separation takes place in a very small area which is the boundary of two electrolyte systems.

Isoelectric electrofocusing (IEF) takes place in a medium in which the pH value of the buffer changes continuously between the electrodes. Amphoteric substances (proteins, peptides) migrate in an electric field towards electrodes with a sign opposite to their charge and stop at the point where the pH value corresponds exactly to the value of the isoelectric point of the separated molecule (pH = pI). At this point, the resultant electric charge of the macro-ion becomes zero so the molecule cannot feel the force of the electric field and stops. In this way, the separation of molecules, mainly proteins, according to their pI values is obtained.

Depending on the needs, these types of electrophoresis can be combined to obtain the desired effect. The combination of zone electrophoresis with isotachophoresis allows for sharper and better separation. The combination of isoelectro-focusing with zone electrophoresis allows a two-dimensional separation (2D electrophoresis) to be performed. The combination of electrophoretic separation with the reaction of specific antibodies allows for the performance of immunoelectrophoresis—rocket immunoelectrophoresis, or affinity immunoelectrophoresis. Depending on the type of chamber in which the electrophoretic separation is carried out, electrophoresis can be divided into capillary, tube, or plate electrophoresis. Agarose or acrylamide gel is most often used as the stationary phase. Electrophoretic separation techniques were used for genome sequencing and made it possible to successfully carry out the Human Genome Project [38]. Examples of applications of electrophoretic methods are described in [39].

3.3 Definition of materiomics

It is obvious that the observed achievements in understanding of both the animate and inanimate world were possible thanks to the joint use of the research techniques described above, as well as research techniques omitted in this study, including information technologies. However, even this cursory review of research techniques makes us realize that the same research methods, slightly modified for different types of research, served purposes in both life sciences and materials science. Despite this, for a very long time the division into separate research areas was effectively maintained, and until today we can observe a reluctant attitude towards the common and mutual use of the achievements in these areas.

This is probably due to the traditions cultivated in the fields they represent, but the lack of basic knowledge of the other fields of knowledge and the lack of willingness to acquire this knowledge also largely influences such a position. Usually, biologists, doctors, and other representatives of the life sciences eagerly use the achievements of materials science, but rarely try to take part in the research process leading to the obtaining of new materials with unprecedented yet but expected properties. Moreover, most often they do not even try to express their expectations regarding such material solutions, while their articulation could be of great importance for directing activities in order to create such expected material. On the other hand, material engineers, chemists, and physicists also rarely use the knowledge of the representatives of the life sciences. Admittedly, it is fashionable to consider solutions developed by nature, which are referred to as biomimetic solutions, but unfortunately often without a sufficiently thorough understanding of the biological processes leading to the creation of a given solution and an explanation of its benefits and limitations of use. It is very likely that the main barrier hindering co-operation is the language used by researchers in both areas, and this is mainly due to the lack or insufficient knowledge of the other area.

It seems that the beginning of the new century has become a turning point in the mutual relations between the life sciences and materials science. In 2003 the team of Tomoki Akita from the Japan National Institute of Advanced Industrial Science and Technology first used the term 'materiomics' during the Combinatorial and Artificial Intelligence Methods in Materials Science Symposium held in Boston, Massachuset, USA. The phrase was literally used in the form: 'The attempt is one of the methods to realize materiomics which is the comprehensive material research based on combinatorial research' [40, 41].

In 2008, in addition to the definition provided by Akita, there came two more definitions. The first was provided by Markus Buehler of the Massachusetts Institute of Technology, USA, as: 'This type of effort, the linking of mechanisms across multiple scales by using a materials science approach to provide structure–process–property links forms the emerging field of materiomics' [42]. On the other hand, a team of researchers from the University of Twente, the Netherlands wrote: 'We propose the name materiomics for the discipline of high throughput methods in biomaterials and tissue engineering science' [43]. It is worth noting that in the original formulation Akita proposed the use of the term materiomics in relation to the general strategy of progress in materials science, Buechler in his definition emphasized the importance of the chain 'structure–process–properties', while in the third definition the concept of materiomics is narrowed down to the part of materials science related to the engineering of biomaterials.

The biomedical sciences successfully use different levels of research, analysis, and understanding (levels of cognition) of biological systems. It is commonly believed that the era known as '-omics' and '-ome' has arrived, taking its name from the common endings of the names of these '-omics' levels of cognition and the common '-ome' endings of accumulated data specific to a given '-omics' level of cognition. A more detailed discussion of the individual selected levels of cognition is available in the review in [44]. Table 3.1 summarizes the most commonly used levels of biological systems of cognition, and the last line describes the level of research, analysis, and understanding in materials science, including the levels of biological systems cognition. In addition to the levels of cognition listed in the table, new ones appear. For example, the collection of all genomes of micro-organisms inhabiting multi-cellular higher organisms, including humans, is called the microbiome, and research in this area is microbiomics. Similarly, the information collected about cellular lipids is called a lipidome, and research in this area is called lipidomics. The issue of nomenclature with the word endings '-omics' and '-ome' was commented on in *The Scientist* [45].

Materiomics can be defined as the holistic, i.e. cross-sectional and integrated, testing of materials, taking into account all available methods of assessing their properties, composition, structure, manufacturing, functionality, methods of safe

Table 3.1. Brief characteristics of the different levels of research, analysis, and understanding (levels of cognition) of biological systems [44]. The last line of the table describes the new level of research, analysis, and understanding in materials science, including the levels of understanding of biological systems.

The level of knowledge '-omics'	The dataset level of knowledge '-ome'	Description of the level of knowledge
Genomics	Genome	Collects all DNA sequences contained in the body.
Transcriptomics	Transcriptome	Collects all the RNA sequences transcribed in the body.
Proteomics	Proteome	Collects all peptides and proteins produced in the body.
Localizomics	Localizome	Collects data on the subcellular location of structures and proteins in the cell.
Interactomics	Interactome	Collects data on the relationship and interactions between macromolecules in the cell.
Metabolomics	Metabolome	Collects and describes all metabolic pathways in the cell and throughout the body.
Materiomics	Materiome	Collects data on the method of material production and describes the relationship between the physicochemical properties, as well as the biological properties/effects of a material and its applications.

use, and processing after use. Materiomics studies the relationships between the methods of production, the functionality of materials and their physicochemical properties, and the effects of their interaction with biological objects. The concept of the materiome should be understood as a set of all available data characterizing a given material, obtained with all available research methods in the field of physicochemistry and biology, both experimental and theoretical or computational, applied at various scales from nano to macro. This is a paradigm similar to the system developed in biology, but applicable to materials science. So far, materiomics has significant applications for, but not limited to, biological materials and biomaterials. Brief information about materiomics was presented by the authors in a study published in the local journal *Laboratorium—Przegląd Ogólnopolski* [46].

3.4 Current applications of materiomics

At present, numerous databases are available collecting the physicochemical properties of materials that often associate them with manufacturing processes. The data collected in these databases are commonly used to select a material due to its physicochemical properties. Similarly, numerous databases characterizing biological objects are available, both at the organism level and at the cellular and molecular levels. This comparison itself again points to an entrenched and still existing division into materials science and life sciences and the resulting difficulties in communication between these disciplines, and in order to make further progress, interdisciplinary action is indispensable. The emergence of the first, and unfortunately the only one so far, database gathering information on biomaterials and the biological effects of their contact with cells, observed at the level of the transcriptome cBiT [47], raises some hope.

Hopefully, the emergence of this first bridging database between biomaterials engineering and transcriptomics is only a prelude to future databases being developed for materiomics and containing a dataset defined as materiome. The cBiT database [48] collects voluntary and fully accessible data on the relationship between the biomaterial, its chemical composition, structure, and the effect in the form of the metabolic response of cells to stress caused by contact with this biomaterial. The measure of the metabolic response of the cell in this case is the change in the gene expression profile, assessed by means of transcriptomics. The relationship between the physicochemical features of the biomaterial and the metabolic response of cells is of great importance in the proper selection of existing biomaterials for biomedical applications, but the possibility of modifying the surface of the biomaterial in such a way as to achieve a predetermined cell response effect has much greater potential. From here, there is only one step to progress and successes in personalized medicine, where it will be possible to develop the best possible surface of biomaterial for a specific patient.

This is somewhat reminiscent of the procedure for selecting an organ donor for the genetic characteristics of the recipient of an organ, however, when using an implant tissue compatibility does not matter, but the specific metabolic response of the cells forming the tissue in contact with the implant is essential. Of course, what the cBiT database offers today is only an introduction to the expectations of researchers both in the field of biomaterials engineering and in biomedical engineering. Since not every change in gene expression translates into a change in the profile of the proteome and other '-omes', it is necessary to further understand the effects of cell contact with the surface of the biomaterial observed at higher levels of cognition and to create new, more comprehensive databases, or creatively modify the former already existing materiome database. It is worth getting acquainted with an extensive work [49] dealing with multilevel hierarchical relations analyzed from the point of view of materiomics.

The literature on the subject includes numerous reports on materiomics in the field of biomaterials engineering and biomedical sciences, where the structures of natural materials, such as bone tissues, filaments, and protein matrices, nucleotide structures and many others, are also analyzed [50–54]. In all of these cases, a materials science approach is used for structural research, and biological techniques are used to study the metabolic response or, more generally, the biological properties of these natural materials. It must be admitted that at present, materiomics has found application mainly in the field of biomaterials engineering, as a special part of materials engineering. However, this idea opens up a further development path also beyond the engineering of biomaterials, particularly in the field of nanomaterials and nanotechnology, due to the far-reaching structural similarities of biologically active biomolecules to engineering nanostructures, as well as the expected or often unexpected biological activity of these synthetic nanostructures [55, 56].

The *International Journal of Nano Research* [57] places materiomics among its leading topics, while the *Journal of Materiomics* [58] aims to provide a continuous forum for the dissemination of research in the general field of materials science, particularly systematic studies of the relationships among the composition, processing, structure, property, and performance of advanced materials.

Working on the border of materials science and life sciences is extremely interesting and fruitful, and mutual enjoyment of the experience of the other field brings only benefits. An example is the molecular engineering of inorganic and hybrid materials [59, 60] and the use of the experience acquired during the implementation of the Human Genome Project, for which it was necessary to process huge amounts of data (big data), which resulted in a new IT strategy for materials science called the Materials Genome Initiative (MGI) [61]. This strategy is entirely consistent with the concept of materiomics, taking into account the collection and processing of large amounts of data stored in databases and thus opening up new possibilities for the development of modern materials, also through a comprehensive analysis of information starting from the atomic and ending with the macroscopic scale [62, 63].

It is also worth mentioning that the observation of the properties of deoxyribonucleic acid (DNA) and its application have resulted in new solutions in the field of nanotechnology, enabling the production of three-dimensional self-organizing nano-objects with controlled shapes [64], and the addition of the possibility of gene editing (CRISPR) [65] almost unlimitedly extends the scope of application of the developed DNA nanotechnologies. At this point, we would like to remind you of our earlier consideration of the super-material that is undoubtedly DNA. These impressive achievements were obtained only thanks to the activity of enthusiasts who can combine the achievements of both fields of science discussed here, but communication difficulties still remain, as evidenced by the fact that there are similar terms in use, but different in nomenclature. The term 'metallomics', proposed in 2018 for the research and description of phenomena caused by metals in contact with biological objects, deserves attention [66].

All this indicates that in the near future there will be further integration of materials science with the life sciences, and the triad of basic sciences, which includes physics, chemistry, and biology, will contribute to a leap in the progress of meeting the material needs of societies, including the need for the production of personalized materials intended for specific people, as well as for specific industrial and economic applications. Already, using the data available in the cBiT database, it is possible to design materials or material surfaces with selected properties in contact with selected cells.

However, this possibility is quite limited, because it concerns those materials and cells for which the tests were carried out and the database contains documented and verified test results. Further development of the cBiT database and the possible creation of further specialized databases in the field of materiomics will significantly accelerate the work and reduce the costs of designing new products for biomedical purposes due to the use of the data collected in this database. However, it should be borne in mind that materiomics is not limited only to work on biomaterials. It can certainly be useful in the work on the safe use and processing of materials. It also seems that it can be useful in the process of identifying engineered materials and nanotechnology products, which was reflected in the patent granted to us [67]. This patent concerns the use of cells as a specific sensor recognizing the surface of the material or nanomaterial particles with which the cell is in contact, and the recognition process can be described as the biological recognition of the stress factor produced by this contact. You will find more on biological recognition in the next chapter of this book.

References

[1] Leewenhoeck A 1684 An abstract of a letter from Mr Anthony Leewenhoeck at Delft, dated Sep. 17 1683. Containing some microscopical observations, about animals in the scurf of the teeth, the substance call'd worms in the nose, the cuticula consisting of scales *Philos. Trans.* **14** 568–74

[2] Parandowski J 1992 *Mitologia. Wierzenia i podania Greków i Rzymian (Mythology. Beliefs and Legends of the Greeks and Romans)* (London: Puls Publiction)

[3] Sorby H C 1858 On the microscopical, structure of crystals, indicating the origin of minerals and rocks *Quart. J. Geol. Soc.* **14** 453–500

[4] Sorby H C 1886 On the application of very high powers to the study of the microscopical structure of steel *J. Iron Steel Inst.* **30** 140–7

[5] Sayre D, Howells M, Kirz J and Rarback H 1987 X-ray microscopy II *Proc. of the Int. Symp., Brookhaven, NY (August 31–September 4)*

[6] Mappes T, Jahr N, Csaki A, Vogler N, Popp J and Fritzsche W 2012 The invention of immersion ultramicroscopy in 1912—the birth of nanotechnology? *Angew. Chem. Int. Ed.* **51** 11208–12

[7] Sroda A, Makowski A, Tenne R, Rossman U, Lubin G, Oron D and Lapkiewicz R 2020 SOFISM: Super-resolution optical fluctuation image scanning microscopy *Optica* **7** 1308–16

[8] Guise O, Strom C and Preschilla N 2011 STEM-in-SEM method for morphology analysis of polymer systems *Polymer* **52** 1278–85

[9] Zhang D, Zhu Y, Liu L, Ying X, Hsiung C-E, Sougrat R, Li K and Han Y 2018 Atomic-resolution transmission electron microscopy of electron beam-sensitive crystalline materials *Science* **359** 675–9

[10] Harris P J F 2018 Transmission electron microscopy of carbon: a brief history *J. Carbon Res.* **4** 4

[11] Gunning W T and Calomeni E P 2000 A brief review of transmission electron microscopy and applications in pathology *J. Histotechnol.* **23** 237–46

[12] Franken L E, Grünewald K, Boekema E J and Stuart M C A 2020 A technical introduction to transmission electron microscopy for soft-matter: imaging, possibilities, choices, and technical developments *Small* **16** 1906198

[13] Smith J W and Chen Q 2020 Liquid-phase electron microscopy imaging of cellular and biomolecular systems *J. Mater. Chem.* B **8** 8490

[14] Shah F A, Ruscsák K and Palmquist A 2019 50 years of scanning electron microscopy of bone—a comprehensive overview of the important discoveries made and insights gained into bone material properties in health, disease, and taphonomy *Bone Res.* **7** 15

[15] King W E, Campbell G H, Frank A, Reed B, Schmerge J F, Siwick B J, Stuart B C and Weber P M 2005 Ultrafast electron microscopy in materials science, biology, and chemistry *J. Appl. Phys* **97** 111101

[16] Transmission electron microscopy vs scanning electron microscopy *ThermoFisher Scientific* https://www.thermofisher.com/pl/en/home/materials-science/learning-center/applications/sem-tem-difference.html

[17] Voigtländer B 2015 *Scanning Probe Microscopy—Atomic Force Microscopy and Scanning Tunneling Microscopy* (Berlin: Springer)
[18] Binnig G and Rohrer H 1982 Scanning tunneling microscope *US Patent* US4343993A
[19] Binnig G and Rohrer H 1987 Scanning tunneling microscopy—from birth to adolescence *Rev. Mod. Phys.* **59** 615–25
[20] Cheng Z, Du S, Guo W, Gao L, Deng Z, Jiang N, Guo H, Tang H and Gao H J 2011 Direct imaging of molecular orbitals of metal phthalocyanines on metal surfaces with an O_2-functionalized tip of a scanning tunneling microscope *Nano Res* **4** 523–30
[21] Binnig G K 1988 Atomic force microscope and method for imaging surfaces with atomc resolution *US Patent* US4724318A
[22] Niedermann P H, Burger J, Binggeu M, Christoph R, Hintermann H E and Marti O 1995 A scanning force and friction microscope *Ultimate Limits of Fabrication and Measurement* ed M E Weiland and J K Gimzewski (Dordrecht: Kluwer Academic) pp 139–45
[23] Cappella B and Dietler G 1999 Force–distance curves by atomic force microscopy *Surf. Sci. Rep.* **34** 1–104
[24] Pimenta-Lopes C, Suay-Corredera C, Velázquez-Carreras D, Sánchez-Ortiz D and Alegre-Cebollada J 2019 Concurrent atomic force spectroscopy *Commun. Phys.* **2** 91
[25] Huey B D, Luria J and Bonnell D A 2019 Scanning probe microscopy in materials science *Springer Handbook of Microscopy* ed P W Hawkes and J C H Spence (Cham: Springer Nature)
[26] Amrein M W and Stamov D 2019 Atomic force microscopy in the life sciences *Springer Handbook of Microscopy* ed P W Hawkes and J C H Spence (Cham: Springer Nature)
[27] Kolodziejczyk A, Jakubowska A, Kucinska M, Wasiak T, Komorowski P, Makowski K and Walkowiak B 2018 Sensing of silver nanoparticles on/in endothelial cells using atomic force spectroscopy *J. Mol. Recognit.* **31** e2723
[28] Kolodziejczyk A M, Kucinska M, Jakubowska A, Sokolowska P, Rosowski M, Tkacz-Szczęsna B, Komorowski P, Makowski K and Walkowiak B 2020 Endothelial cell aging detection by means of atomic force spectroscopy *J. Mol. Recognit.* **33** e2853
[29] Stoica P and Moses R 2005 *Spectral Analysis of Signals* (Upper Saddle River, NJ: Prentice Hall)
[30] Pedrotti L 2008 *Basics of Spectroscopy* (Orlando, FL: National Center for Optics and Photonics Education)
[31] Fleming I and Williams D 2019 *Spectroscopic Methods in Organic Chemistry* 7th edn (Cham: Springer Nature)
[32] Gerothanassis I P, Troganis A, Exarchou V and Barbarossou K 2002 Nuclear magnetic resonance (NMR) spectroscopy: basic principles and phenomena, and their applications to chemistry, biology and medicine *Chem. Educ.: Res. Pract. Eur.* **3** 229–52
[33] Murphy D M 2009 EPR (Electron paramagnetic resonance) spectroscopy of polycrystalline oxide systems *Metal Oxide Catalysis* ed S D Jackson and J S J Hargreaves (Weinheim: Wiley)
[34] de Hoffmann E and Stroobant V 2007 *Mass Spectrometry Principles and Applications* 3rd edn (Chichester: Wiley)
[35] Coskun O 2016 Separation techniques: chromatography *North Clin. Istanb.* **3** 156–60
[36] Walkowiak B 2000 *Techniki chromatografii cieczowej. Przykłady zastosowań (Liquid Chromatography Techniques. Examples of Applications)* (Lublin: Amersham Pharmacia Biotech) http://www.biofizyka.p.lodz.pl/chtytul.pdf

[37] Magdeldin S (ed) 2012 *Gel Electrophoresis—Principles and Basics* (Rijeka: InTech)
[38] *Human Genome Project Information Archive 1990–2003* https://web.ornl.gov/sci/techresources/Human_Genome/project/index.shtml
[39] Walkowiaka B and Kochmańskiej V 2002 *Elektroforeza. Przykłady zastosowań. Oparcowanie zbiorowe pod edycją (Electrophoresis. Examples of Applications. Collective Study)* (Amersham Biosciences, local office in Warsaw, Poland) http://www.biofizyka.p.lodz.pl/elektroforeza.pdf
[40] Akita T, Udea A, Yamada Y, Ichikawa S, Tanaka K, Kohyama M and Kobayashi T 2003 Analytical team observations of combinatorial catalyst libraries for hydrogen production—as a part of materiomics *Symp. Combinatorial and Artificial Intelligence Methods in Materials Science II (Boston, MA, 1–4 December)* JJ9.5
[41] Yamada Y *et al* 2003 Combinatorial catalysis survey concerning proton exchange membrane fuel cell technology—as a part of MATERIOMICS *Symp. Combinatorial and Artificial Intelligence Methods in Materials Science II (Boston, MA, 1–4 December)* JJ9.4
[42] Buehler M J, Keten S and Ackbarow T 2008 Theoretical and computational hierarchical nano mechanics of protein materials: deformation and fracture *Prog. Mater. Sci.* **53** 1101–241
[43] Van Blitterswijk C, Stamatialis D, Unadkat H, Papenburg B, Rouwkema J, Truckenmüller R, Van Apeldoorn A, Wessling M and De Boer J 2008 Materiomics: dealing with complexity in tissue engineering *Tissue Eng.* A **14** 796
[44] Mackiewicz P, Zakrzewska-Czerwińska J and Cebrat S 2005 Genomika—dziedzina wiedzy XXI wieku (Genomics—a field of knowledge in the 21st century) *Biotechnologia* **3** 7–21
[45] Lederberg J and McCray A 2001 Ome sweet omics—a genealogical treasury of words *Scientist* **15** 8
[46] Walkowiak B, Walkowiak-Przybyło M and Komorowski P 2019 Materiomika—pomost łączący nauki o materiałach z naukami życiu (Materiomics—a bridge connecting materials science with life sciences) *Laboratorium—Przegląd Ogólnopolski* **2** 24–30
[47] Hebels D G A J, Carlier A, Coonen M L J, Theunissen D H and de Boer J 2017 cBiT: a transcriptomics database for innovative biomaterial engineering *Biomaterials* **149** 88–97
[48] *cBiT Database*
[49] Cranford S W and Buehler M J 2012 *BiomateriomicsSpringer Series on Materials Science* vol 165 (Berlin: Springer)
[50] Cranford S and Buehler M J 2010 Materiomics: biological protein materials, from nano to macro *Nanotechnol. Sci. Appl.* **3** 127–48
[51] Buehler M J and Ballarini R (ed) 2013 *Materiomics: Multiscale Mechanics of Biological Materials and Structures* vol 546 (Berlin: Springer)
[52] Cranford S W, de Boer J, van Blitterswijk C and Buehler M J 2013 Materiomics: an -omics approach to biomaterials research *Adv. Mater.* **25** 802–24
[53] Launey M E, Buehler M J and Ritchie R O 2010 On the mechanistic origins of toughness in bone *Annu. Rev. Mater. Res.* **40** 25–53
[54] Groen N, Guvendiren M, Rabitzc H, Welshd W J, Kohnb J and de Boer J 2016 Stepping into the omics era: opportunities and challenges for biomaterials science and engineering *Acta Biomater.* **1** 133–42
[55] Reisner D E, Bawa R, Brauer S, Alvelo J, Zheng W, Gericke M and Vulpe C 2012 Bionanotechnology *Handbook of Research on Biomedical Engineering Education and Advanced Bioengineering Learning: Interdisciplinary Concepts* (Hershey, PA: IGI Global)

[56] Hulshof F F B, Zhao Y, Vasilevich A, Beijer N R M, de Boer M, Papenburg B J, van Blitterswijk C, Stamatialis D and Boer J D 2017 NanoTopoChip: high-throughput nano-topographical cell instruction *Acta Biomater.* **62** 188–98

[57] *International Journal Nano Research* https://www.innovationinfo.org/international-journal-of-nano-research/aimsandscope

[58] *Journal of Materiomics* https://www.journals.elsevier.com/journal-of-materiomics/

[59] Sanchez C *et al* 2014 Molecular engineering of functional inorganic and hybrid materials *Chem. Mater.* **26** 221–38

[60] Saveleva M S, Eftekhari K, Abalymov A, Douglas T E L, Volodkin D, Parakhonskiy B V and Skirtach A G Hierarchy of hybrid materials—the place of inorganics-in-organics in it, their composition and applications *Front. Chem.* **7** 179

[61] Agrawala A and Choudhary A 2016 Perspective: materials informatics and big data: realization of the 'fourth paradigm' of science in materials science *APL Mater.* **4** 053208

[62] Hill J, Mulholland G, Persson K, Seshadri R, Wolverton C and Meredig B 2016 Materials science with large-scale data and informatics: unlocking new opportunities *Mater. Res. Soc. Bull.* **41** 339–409

[63] Rickman J M, Lookman T and Kalinin S V 2019 Materials informatics: from the atomic-level to the continuum *Acta Mater.* **168** 473–510

[64] Douglas S M, Dietz H, Liedl T, Högberg B, Graf F and Shih W M 2009 Self-assembly of DNA into nanoscale three-dimensional shapes *Nat. Lett.* **459** 414–8

[65] Doudna J A and Charpentier E 2014 The new frontier of genome engineering with CRISPR-Cas9 *Sci. Rev.* **346** 1258096

[66] Maret W 2018 Metallomics: the science of biometals and biometaloids *Metallomics: The Science of BiometalsAdvances in Experimental Medicine and Biology* ed M A Z Arruda (Berlin: Springer)

[67] Walkowiak B, Komorowski P and Walkowiak-Przybyło M 2012 Sposób identyfikacji produktów inżynierii materiałowej lub nanotechnologii (Means of identification of materials engineering products or nanotechnologies) *Polish Patent* Pat.223224, P.399899

IOP Publishing

Biological Evaluation of Materials
The interaction of materials with their environment

Bogdan Walkowiak, Magdalena Walkowiak-Przybyło and Piotr Komorowski

Chapter 4

Biological recognition of materials

Progress in science is driven by the acquisition of new information about the world around us, which inevitably leads to the need to revise well-established scientific views and dogmas by introducing new ideas and laws that better describe reality. However, the emergence of new theories is limited by the principle of correspondence in force in science, which imposes the necessity of the correspondence of the new theory with the theories previously in force. On the basis of such understanding of development, it turned out to be possible to create a synthetic cell, as well as to explain the processes of molecular interactions and molecular recognition which lie at the basis of chemical technologies and biochemical processes. This chapter also discusses the use of molecular biology methods to recognize stress factors by using a cell, which can be used to identify materials.

4.1 Life and matter

The extraordinary richness and variety of life forms evoke justified admiration, but also prompts reflection and deliberations on the mechanisms that enable life to last, often against unfavorable environmental conditions which force living organisms to adapt to these conditions. With the passage of time and the accumulation of experience and scientific knowledge, the way we understand life has changed. In the distant past, the primitive rational man hardly thought about the phenomenon of life, but with the deepening of his awareness of his own existence, a thought appeared separating the animate world—endowed with life—from the inanimate material world.

With the emergence of awareness of the existence of life, the question of how life originated and the mechanisms of its duration naturally arose. The problem of explaining the way life originated largely belonged to the religious–philosophical category. Until the nineteenth century, that is until the theory of evolution of organisms was developed, there was a common view of the creation of life on Earth by God [1] or deities, known as creationism [2]. As science developed and

doi:10.1088/978-0-7503-2656-8ch4

accumulated knowledge about living organisms and their evolutionary variability, as documented by Charles Darwin (1809–1882) [3], the idea known as the theory of evolution became widely accepted and modified with the emergence of new scientific facts [4]. Creationists do not agree with this idea, as they present their modified views on the origin of life and its changeability in the form of intelligent design [5].

Without offending anyone's religious feelings, our aim is not to discuss the origins of life and the sources of the driving forces of this life along with its changeability. Rather, we want to continue our earlier story about the evolution of the Universe based on the scientifically documented and derived laws of physics and chemistry, which are the basis for the description of the processes taking place both in the area of inanimate matter and in relation to living matter. The basis for treating the subject of our considerations in this way is the fact that the successes of physics and chemistry in describing the world around us are based on the simple fact that all matter, both animate and inanimate, consists of the same atoms, which in turn contain the same components themselves in the form of electrons and nucleons, and the nucleons in turn contain the same quarks, whether the atoms are in living or inanimate matter.

For the purposes of our deliberations on animate matter and life, we can assume that life is a set of highly functionally organized processes, physical transformations, and chemical reactions, taking place in systems that are thermodynamically open but separated from the environment, always containing nucleic acids, proteins, lipids, and oligosaccharides. Living objects are characterized by the ability to reproduce and develop changes that are beneficial for the organism. In this sense, viruses that contain nucleic acids and proteins but are unable to reproduce themselves are not living objects. We usually assume that life takes place on Earth, but do not exclude its presence on other planets in the Universe. In such a definition of life, all life processes require the presence of organic compounds based on carbon chemistry, which does not mean that an alternative life based on, for example, silicon is not possible in the Universe, although in the light of knowledge and experience in the field of organic chemistry, the functioning of organosilicon compounds that might create alternative life form is highly problematic [6].

Despite the conducted radio astronomy surveys, so far it has not been possible to detect signals that could indicate the presence of other intelligent life in the Universe, but there is abundant and well-documented information about the presence of complex organic substances both in the Solar System and in distant galaxies [7]. From this fact, it should be concluded that the organic compounds necessary for the emergence and functioning of life may arise and condense spontaneously throughout the cosmos, which may or may not prove the existence of life beyond Earth. This issue is still unclear and is the subject of lively scientific discussions [8].

Thanks to the successes of biology, it was possible to delve into the structure of living organisms, including the isolation of specialized tissues and the cells that compose them. In further stages of cognition, specialized macromolecules capable of performing highly specific functions were distinguished. Initially, proteins were indicated as the most important for the functioning of the organism. This was also reflected in philosophical considerations concerning life. Fryderyk Engels

(1820–1895), on the basis of the materialism he represents and as a continuation of Georg Wilhelm Fryderyk Hegel (1770–1831) ideas of life as a self-renewing chemical process that lasts forever (life is a self-renewing chemical process made perennial), expressed the belief that life is the mode of existence of proteins [9]. This statement gained considerable popularity and it is worth mentioning that in 1970 a song appeared in the repertoire of the Polish rock group Skaldowie with the phrase 'life is a form of protein existence' in the chorus [10]. The song was written by Agnieszka Osiecka (1936–1997). Regardless of the popularity of this statement, it contains a very significant dose of the scientifically proven truth that proteins are necessary for the functioning of life and active participation in biochemical processes taking place in the cell, but also for the control of these processes.

Unfortunately, this statement does not fully exhaust the complexity of the molecular mechanisms involved in the cell. In particular, nucleic acids play an extremely important role in both storing, sharing, and transferring the genetic information of a cell. However, proteins are involved in all processes related to the use of genetic information in a cell, i.e. reading and transferring information from DNA to RNA (transcription), DNA duplication (replication), and the synthesis of peptides on the RNA template (translation). Proteins are also responsible for controlling the integrity and complementarity of DNA strands and repairing any damage that occurs, as well as post-translational modification of the synthesized peptides and assembling them into biologically active proteins. However, the role of proteins is not limited to these processes.

The whole metabolism of the cell is based on the activity of specialized proteins, including contact with the external environment and selective transport of substances through the cell membrane and intracellular membranes, mainly composed of phospholipids, but also containing membrane proteins. Structural cytoskeletal proteins are responsible for the shape of the cell and its changes, which also take place due to mechanisms that use protein activity. Proteins are also involved in labeling other proteins for enzymatic degradation. After all, proteins are also involved in the control of cell adaptation processes responding to changes in environmental conditions and in the control of the functional state of the cell, and in the event of damage that cannot be repaired, the proteins trigger the process of programmed cell death—apoptosis.

Often, the role of polysaccharides is overlooked in considerations about the functioning of the cell. This has its historical condition, as for a long time the role of polysaccharides has been diminished, and molecular biology has been reduced only to the biochemistry of nucleic acids and proteins. For this reason, the role of polysaccharides is still not fully understood and appreciated. It should be noted, however, that glycoproteins and glycolipids take part in cellular processes, and free glucose, which appears in excess in diabetes mellitus, impoverishes and disturbs biochemical processes through non-specific glycation of proteins. As can be seen from the above short overview, proteins interacting with nucleic acids, phospholipids, and polysaccharides are an essential functional element of cells, but also of the extracellular matrix and body fluids, and in this sense the form of protein existence strongly defines life.

In order to broaden our knowledge of the processes taking place in the cell, it is worth getting acquainted with biochemistry textbooks, including one of the first, published for the first time in 1953 [11], although, due to the intensive progress in knowledge about life at the cellular and molecular level, the reader is advised to have a closer look at more recent books, including the publicly available Harper textbook [12].

4.2 The principle of correspondence in science

As the history of science teaches, the accumulation of new knowledge sooner or later leads to some inconsistencies between generally accepted theories and emerging observations. Initially, attempts are made to repeat the experiments in order to exclude observational errors, but after confirming the deviation of the obtained results from the predictions of the current theory, a breakthrough occurs and another theory emerges that explains these observed deviations well, as well as all the results of experimental research to date. The latter statement includes the principle of correspondence that is in force in science. This principle applies to the relationship between two theories, one of which is an extension or generalization of the other previously valid theory. The new theory must explain all the observations explained so far by the old theory. In the mathematical description of phenomena, it is usually required that the equations formulated for a new theory, under certain assumptions, could be reduced to the equations of the previously valid theory. The best recognizable example is the heliocentric theory, placing the Sun instead of the Earth at the center of the world known at the time, formulated by Nicolaus Copernicus (1473–1543) and presented in the work 'On the revolutions of the heavenly spheres' (De revolutionibus orbium coelestium), which replaced the previously valid geocentric theory, according to which the Earth was in the center of this world and all celestial bodies rotated around it. Of course, in the light of our present knowledge, the Sun lies at the edge of the Milky Way and is by no means the center of the world.

A more recent example is the theory of special relativity formulated by Albert Einstein (1879–1955), whose equations transform into those of classical physics under conditions of low speed of motion. The same is true of Einstein's general theory of relativity, which generalizes the special theory of relativity, valid only for inertial frames of reference, to all frames of reference, and is also an extension of the classical theory of gravity formulated by Isaac Newton (1643–1727). The heliocentric theory and both theories of relativity, along with quantum theory, have become a turning point in physics and are among those breakthrough theories that are generally known, though not always well understood by the general public.

An example of similar importance in the field of life sciences, but less recognizable than the examples above, is certainly the central dogma of molecular biology explaining the flow of genetic information in biological systems. This dogma was formulated by Francis Crick (1916–2004) in 1957 [13] with the following statement (reproduced with permission of Wellcome Collection): '...once 'information' has passed into protein it cannot get out again. In more detail, the transfer of information from nucleic acid to nucleic acid, or from nucleic acid to protein may be possible, but transfer from protein to protein, or from protein to nucleic acid is impossible. Information means here the precise determination of sequence, either of bases in the nucleic acid or of amino acid residues in the protein.'

This dogma often exists in a simplified form proposed by James Watson (1928–), co-author, with Crick, of the concept of the double-stranded structure of DNA. The simplified version describes the two-stage process of information flow with respect to the direction from DNA to RNA and then from RNA to protein [14]. This simplified version of the dogma was refuted with the discovery of the existence of RNA viruses (retroviruses) and the description of the activity of reverse transcriptase capable of transcribing information from RNA into DNA [15]. Currently, this process is commonly used in research laboratories to synthesize RNA copies in the form of complementary DNA using the polymerase chain reaction (PCR) technique.

Unfortunately, in science one should be very careful with dogmas, as they are usually not very durable and it should be taken into account that the central dogma of molecular biology will also be at least modified [16]. The results of research and observations show that it is possible to transfer information about the spatial structure of a protein directly from one protein molecule to another. An example of such a process is the presence of pathological forms of the prion protein capable of infecting normal forms of this protein by contact and transforming them into a pathological form without the apparent involvement of any other infectious agent and DNA or RNA participation [17], which results in the incurable Creutzfeldt–Jakob disease (CJD) in humans [18]. However, in opposition to the prion explanation of the causes of CJD, there is a paper describing the participation of a small, very mobile, and devoid of cell walls *Spiroplasma* bacterium in the infection of animal counterparts of prion disease observed in ruminants [19], as well as another review publication based on the results of works by other authors [20]. The disadvantage of this concept is the lack of confirmation by other research teams, which goes hand in hand with the widespread recognition of the prion causes of CJD.

Regardless of how the fate of explaining the causes and course of CJD and the possibility of transferring information from one protein to other proteins, this situation, and the previously described relationships between old and new theories, show us two important facts. First, in science we should not place any particular faith in dogmas and should be open to new research results, not always consistent with predictions formulated on the basis of existing and generally accepted theories. Second, as it follows from the considerations on the central dogma of molecular biology, the intermolecular interactions are important in the phenomenon of life, regardless of the direction of information transfer. It is thanks to these interactions that life is possible and lasts. For the phenomenon of life, the mere existence of proteins, nucleic acids, sugars, and fatty acids is not enough, the essence of life is their mutually ordered and controlled interactions as a result of which the necessary life processes are realized.

4.3 Artificial cell

Since the identification of the cell as a living organism, or a component of a living multicellular organism, there has been a desire to understand the mechanisms of cell functioning at the molecular level, which means identifying the metabolic processes and pathways responsible for the proper functioning of the cell. Advances in the area of cell biochemistry allowed for the identification of numerous pathways and metabolic cycles, as well as the direction of information flow in the cell, but it still did not provide a full answer to the basic question about the organization and management of these processes, the more so as the numerous genes present in the genetic material of the cell seemed to be unused, even redundant.

Obviously, the question arose about the minimum requirements for a cell in terms of gene abundance and diversity. In order to answer this question, the researchers' attention was drawn to a very simple and small bacterium without a cell wall called *Mycoplasma*, which, due to its minimal vital functions, was suspected of having the smallest number of functional genes. Studies carried out on the *Mycoplasma genitalium* showed that the complete nucleotide sequence encompassed 580 070 base pairs of the organism's genome, which allowed the identification of only 470 coding regions including genes required for DNA replication, transcription and translation, DNA repair, cellular transport, and energy metabolism [21]. Comparison of this genome with that of other simple micro-organisms suggests that differences in genome content reflect profound differences in the physiology and metabolic capacity of these organisms. In order to identify the minimum set of genes necessary for the life of the cell, further studies were performed to remove each gene sequentially and tested the survival of the resulting mutants, which allowed the minimum set of 380 genes necessary for life to be identified [22].

In 2010 the complete genome was chemically synthesized based on the sequence of the genome of the *Mycoplasma mycoides* cell and it was transplanted into another *Mycoplasma capricolum* cell, from which the original DNA had been previously removed [23]. This is how a new bacterium was created that was still viable. It was named *JCVI-syn1.0* or *Synthia*. Further experiments allowed the smallest set of genes sufficient for the functioning of the organism to be identified and a cell called *JCVI-syn3.0* to be created, containing 473 genes, 149 of which have an unknown function so far [24]. This cell was considered to be the first synthetic organism, although the term synthetic strictly refers to the method of obtaining the cell's genome, and does not correspond to the rest of the cell components that came from another living cell and arose naturally. The creation of a completely synthetic organism still remains a task to be performed with appropriately developed strategies [25–27]. Methods of controlling the shape and size of synthetic cells [28], the possibility of self-reproduction of membranes [29], building a chromosome [30], transcription and translation processes [31], as well as genetic requirements for cell division [32] have already been developed.

These achievements, although they seem like fantasy, are completely real and repeatable. They are a tremendous achievement in molecular engineering. A synthetic cell can function by ensuring the existence of the necessary macromolecules and the possibility of entering them into controlled molecular interactions, with this control being exercised by protein macromolecules found in the cell and synthesized in it as needed. This indicates the invaluable role of molecular interactions in the life processes of organisms, regardless of their complexity.

All the forces occurring between molecules are responsible for the molecular interactions, except for the customary excluded covalent bonds, which are characterized by much greater strength and durability. These are, therefore, ion–ion interactions, ion–dipole interactions, dipole–dipole interactions, permanent dipole–induced dipole interactions (van der Waals interactions), induced dipole–induced dipole interactions (London dispersion interactions), hydrogen bond interactions—resulting from sharing a partial positive charge of hydrogen with a partial negative charge of another molecule—, as well as the repulsive interaction between electrons in atoms in close proximity to each other. All these forces contribute significantly to the maintenance of specific spatial structures by macromolecules as a result of interactions within molecules and interactions with the surrounding environment, which in the case of living organisms are polar water molecules. A stable conformational state of a molecule is achieved when the energy minimum of the molecule is reached, but this minimum may be a local minimum and as a result of energy exchange (obtaining and returning) the molecule may go to another local minimum or reach the global minimum guaranteeing the highest conformational stability, usually without proper biological activity, often showing pathological properties as in the case of amyloid fibers [33]. Many biologically active macromolecules, after synthesis in the cell, take one of the possible local minima, determining proper activity, thanks to appropriate folding under the control of chaperones [34]. It is also one of the most serious challenges of the pharmaceutical industry in the production of recombinant protein drugs synthesized in cells of other organisms, where they are not affected by the proper chaperones [35].

4.4 Molecular interactions and recognition

The issues of molecular interactions are dealt with in particular by supramolecular chemistry [36]. Historically, in 1873 Johannes Diderik van der Waals (1837–1923) proposed the existence of intermolecular interactions, but the first concept of supramolecular chemistry was proposed in 1890 by Hermann Emil Fischer (1852–1919) as the 'lock and key' mechanism to explain substrate recognition by the enzyme. This pictorial picture explaining the specificity of enzymatic reactions is still used today and has been developed into a host–guest mechanism describing molecular complexes consisting of two or more molecules or ions held in a unique structure by non-covalent interactions [37].

With the discovery of hydrogen bonds in the twentieth century, there was a significant acceleration in the area of molecular interaction studies, which allowed for understanding the internal and spatial structures of proteins, nucleic acids, polysaccharides, and fatty acids, as well as many other structures, sporadically or not found in living cells. Various concepts of interaction are considered within supramolecular chemistry. Among other things, the following are considered: problems of molecular self-organization enabling the adoption of a biologically active spatial conformation [38, 39]; molecular recognition and resulting complex formation [40, 41]; template synthesis, examples of which are transcription—RNA synthesis on a DNA template and translation—peptide synthesis on an mRNA template [42, 43]; molecular marking (imprinting), examples of which are postranslational modifications of proteins (e.g. glycation, phosphorylation, ubiquitination) [44], or DNA methylation [45]; the action of molecular machines, such as the muscle [46] or the reshaping and movement of cells [47]; breaking covalent bonds (dissociation) and the reverse process of formation of covalent bonds under the control of non-covalent forces to create a structure of lower energy, exemplified by the activity of proteolytic enzymes and peptide synthesis as well as enzymes digesting and synthesizing nucleic acids [48, 49]. Supramolecular interactions are also responsible for the formation of mechanically locked structures consisting of molecules linked with each other due to their topology [50]. Understanding the phenomena of molecular interactions brought spectacular results in the form of synthetic biomimetic structures finding practical applications [51–53].

'Nothing can happen in biology unless something binds to something else'—this is a very momentous quote [54] (reproduced with permission of Wiley). For the proper functioning of a living organism and controlled contacts with the environment, molecular recognition is of particular importance, which determines both the control of biochemical processes inside the cell, including the transduction of intracellular signals [55], and the transduction of signals propagated outside the cell (release of signal molecules and obtaining information through specific binding of signaling molecules by receptors) [56, 57].

Moreover, the effectiveness of molecular recognition determines the proper functioning of the immune system response of vertebrates, including humans, to the emergence of threats from viruses, bacteria, fungi, or other pathogens [58, 59], as well as the development of neoplastic diseases [60]. Molecular recognition underlies the mechanism of recognizing antigens as self or foreign, enabling the immune system to control the histocompatibility [61]. The same mechanism is involved in the process of developing immunity as a result of natural immunization, which is the disease from infection, as well as by immunization with vaccines. There is no doubt that thanks to vaccinations, the incidence of serious infectious diseases has been greatly reduced or completely eliminated [62]. It is also of great importance in the context of the ongoing COVID-19 pandemic caused by the SARS-CoV-2 virus [63]. In the vast majority of cases, we observe a transient formation of complexes, which, however, plays a decisive role in the processes of molecular recognition [64, 65] triggering all physiological processes in the body.

The phenomenon of molecular recognition is also of practical importance in many analytical and preparative applications, including the detection of factors posing a threat to living organisms [66], affinity chromatography (AC) [67], diagnostic enzyme immunoassays (ELISA), radioimmunoassays (RIA), immunofluorescence tests (IFA), and many other serological tests [68], observations of the formation and disintegration of molecular complexes in real time using the surface plasmon resonance (SPR) technique [69], assessment of thermodynamic complexation reaction parameters in microcalorimetric tests (ITC) [70], and many other practical applications. Most of the above-mentioned tests allow the evaluation of the binding strength of the complexes expressed by the value of the association constant, and thus the differentiation of the affinity of the complex components [71–73].

It can be assumed that both the cell and the multicellular organism are able to use molecular recognition in order to recognize the stress resulting from changes in the environment or from exposure to contact with the new environment and to initiate adaptive processes [74, 75], and the observed effect is called metabolic response to stress factors [76, 77] and includes changes in gene expression [78, 79].

Molecular recognition is also successfully used for the detection and identification of materials engineering and nanotechnology products, and the detection process itself, due to the use of biological objects in it, is sometimes referred to as biological recognition. For example, graphene flakes are coated with serum proteins in a specific and reproducible way, which, using specific antibodies, allows for their detection and characterization of their dispersion [80].

Another example would be the use of a multicellular model organism, the nematode *Caenorhabditis elegans* (*C. elegans*), used to detect the presence of certain nanoparticles. It turns out that exposure of this nematode to silver nanoparticles, multi-walled carbon nanotubes, and fourth generation PAMAM dendrimers, even in very low concentrations, causes significant and specific changes in gene expression, but also measurable population changes, which can be used for early detection of the presence of these nanoparticles [81] to alert on potential dangers. This property of *C. elegans* organism was used to create a functional sensor capable of detecting the presence of silver nanoparticles and assessing its toxicity [82].

4.5 Signal transduction and particles uptake

Let us try to trace some possible mechanisms of the interaction of nanostructures with biological objects and indicate the possible effects of this. Figure 4.1 shows

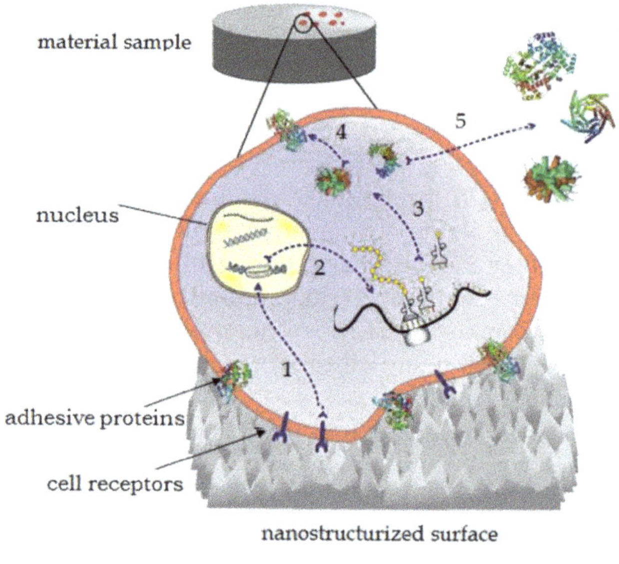

Figure 4.1. Schematic representation of selected metabolic processes in eukaryotic cell exposed to contact with a nanostructured surface. 1—transduction of a signal that triggers a change in gene expression; 2—release of mRNA molecules transcribed on the template of the active gene; 3—synthesis of peptides in ribosomes on an mRNA template (translation process) and controlled folding and post-translational modification of proteins; 4—distribution of active proteins to the localization of their activity (different cell compartments, cytosol, cell membrane); 5—release of extracellular matrix proteins and proteins involved in signal transduction outside the cell.

schematically selected hypothetical metabolic response mechanisms triggered in the cell as a result of contact with a nanostructured surface (it may be the surface of a macroscopic material sample or the surface of a nanoparticle that has not penetrated into the cell interior).

The key conclusion from the analysis of the diagram in figure 4.1 is the activation of the metabolic response of the cell as a result of the stress arising from its contact with the nanostructured surface. It should be noted that the surface of each macroscopic object analyzed on a nanometric scale presents its nanostructure inaccessible to our senses, but accessible to receptors and membrane structures present on the cell surface. Therefore, even the same macroscopic objects can induce a diverse metabolic response of cells in contact with them, depending on the surface condition of this object defined in the nano-scale (presence of different molecules adsorbed on the surface as a result of a different history of surface contact with a changing biological environment). In the case of particles capable of penetrating the cell membrane, the process of causing changes in gene expression may take place already at the moment of contact with the cell membrane or later as a result of contact with the internal structures of the cell, including numerous proteins. As it is easy to notice during the analysis of the diagram in figure 4.1, the expected effect of the contact of the cell with the artificial surface is, in the first place, the transduction of signals that trigger changes in gene expression, and thus also changes in the profile of gene expression products, i.e. peptides and proteins.

The surprisingly orderly nature of these interactions can be seen in figure 4.2, where cells exposed to contact with microdiamond particles are presented (the average size of diamond crystals was about 1 μm). The grouping of nanoparticles surrounding the convexity of the cell nucleus is clearly visible, while the convexity itself and the extracellular space are mostly free of particles. It is worth paying attention to the fact that diamond particles do not penetrate the bone cells (osteoblasts; Saos-2 cell line, figures 4.2(a) and (b)), while at the same 24 h of exposure, diamond particles penetrate into the interior of the vascular endothelial cells in significant numbers (cell line EA.hy926, figures 4.2(c) and (d)). In another of our works we showed that diamond particles are similarly accumulated and taken up by epithelial cells of the small intestine (FHs74Int cells) [83]. We also observed an analogous effect of the accumulation and uptake of silver nanoparticles by endothelial cells forming the blood–brain barrier (HBEC5i cells) [84], and in yet another of our work we showed that endothelial cells are prone to similarly absorb other nanoparticles, including multi-walled carbon nanotubes (MWCNT) and polyamidoamine dendrimers (PAMAM) [85].

Specialized barrier cells (epithelium and endothelium), whose task is to selectively pass (uptake and release) substances, are definitely more susceptible to the process of absorption of diamond particles and various nanoparticles than other cells. It is obvious that such specialized barrier cells should exhibit some common characteristics, including a high capacity to develop endocytosis, but the metabolic response of these cells is only partially similar and that they largely exhibit a highly specific metabolic response.

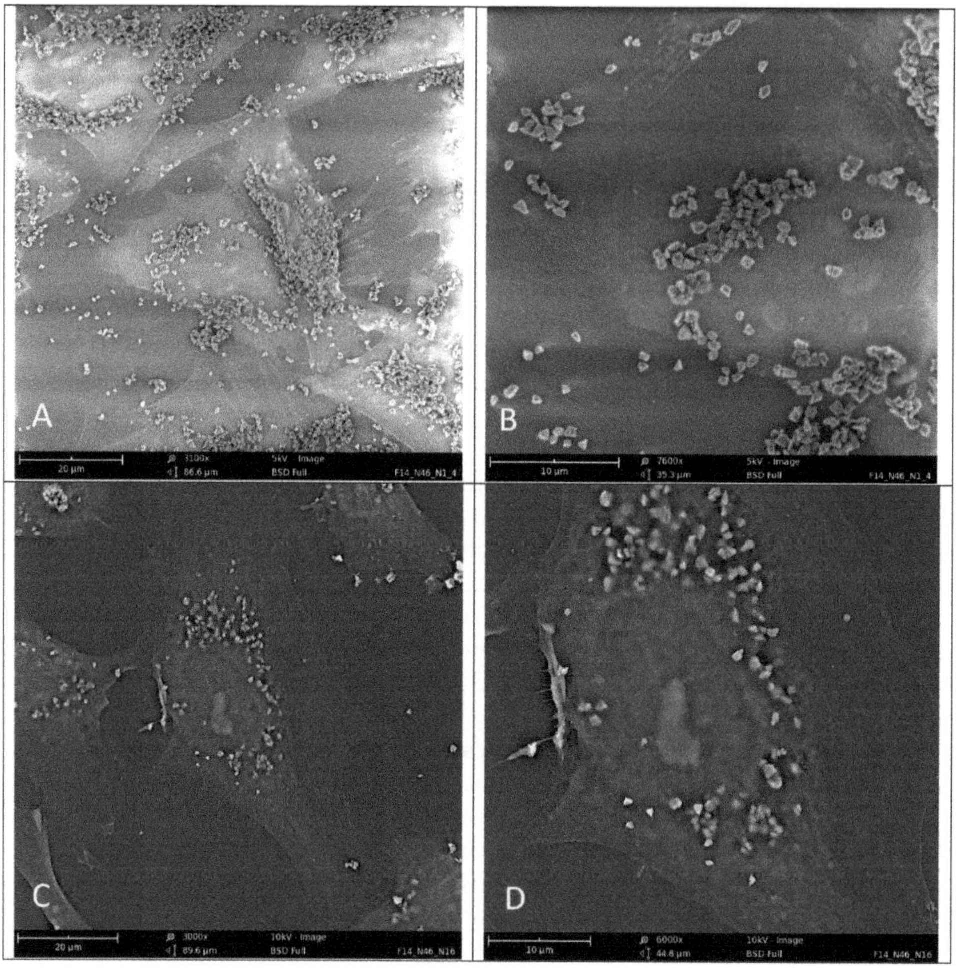

Figure 4.2. Microdiament particles in contact with osteoblasts (Saos-2) and with vascular endothelial cells (EA.hay926). In the case of endothelial cells, the penetration of diamond particles into the interior of the cell is clearly visible.

4.6 Molecular biology methods

The resources of molecular biology research methods include techniques that track changes in gene expression (transcriptome) as well as changes in the protein profile of a cell (proteome). Below is a brief description of the methodology used to track changes in the transcriptome and proteome profiles of a cell. Figure 4.3 shows a schematic view of the experiment that allows for the observation of changes in gene expression (transcripts).

In an exemplary transcriptomic experiment, two samples of the same cells were grown under the same optimal conditions except for the presence of nanoparticles (case 1) or their absence (case 2—control). After completion of culture, cells were lysed and the entire pool of RNA molecules was isolated. Then the number of RNA

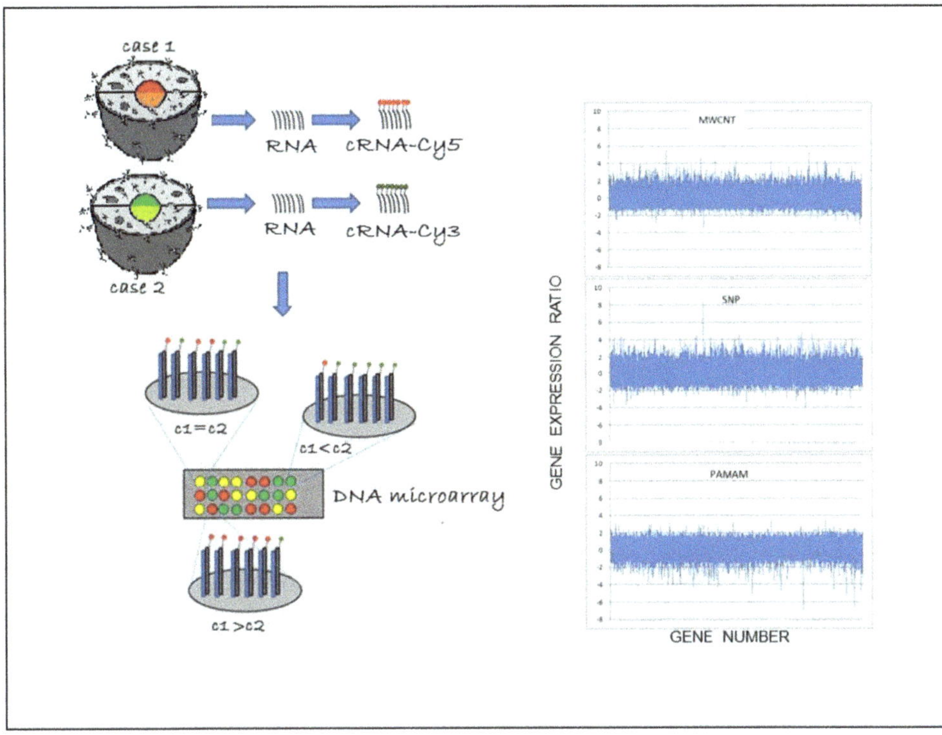

Figure 4.3. Schematic presentation of an experiment to assess changes in gene expression (transcriptome) of cells in response to stress (contact with multi-walled carbon nanotubes (MWCNTs), nanosilver particles (SNPs), and PAMAM dendrimers). Explanations in the text below.

molecules was multiplied and fluorescent markers Cy3 and Cy5 (respectively) were incorporated using the polymerase chain reaction (PCR [86]) and the obtained products (molecular probes) in the form of fluorescently labeled RNA copies (cRNACy5—case 1; cRNA-Cy3—case 2) were used in the process of hybridization (molecular recognition and complexation) with DNA fragments representing particular genes on the DNA matrix [87].

After hybridization and washing of the microarrays (removal of unspecifically adsorbed molecular probes) the microarrays were scanned using a dedicated high-resolution fluorescence scanner. As a result, a colored image of spots was obtained, where spots of genes with the same expression in both cell samples are marked in yellow (active genes with the same expression level in both samples, $c1 = c2$), spots containing genes in cells exposed to nanoparticles (case 1) in increased expression (overexpression) compared to control cells are marked red ($c1 > c2$), while spots representing genes with reduced expression (suppression) in the tested cells in comparison to control cells are marked in green ($c1 < c2$). Fluorescence intensity analysis allows for the quantification of the expression of individual genes in cells and the presentation of the results in the form of a ratio of gene expression levels in the test cells in relation to the control cells (fold change—FC). The result of this particular experiment is presented in the form of three separate graphs (figure 4.3). At first glance, these plots are very similar, but a more in-depth analysis shows those genes whose expression is clearly altered both towards overexpression and suppression. At the same time, it is obvious that the pattern of changes in gene expression is different for each of the studied nanomaterials.

Figure 4.4 shows a schematic view of the experiment that allows the observation of changes in the peptide profile (proteome).

In an exemplary proteomics experiment, samples of the same cells were grown under the same optimal conditions except for the presence of nanoparticles (test sample) and their absence (control sample). After completion of the culture, the cells were lysed in the presence of CHAPS detergent, urea and thiourea, as well as a disulfide reducing agent (dithiothreitol—DDT) and ampholytes (IPG-buffer). The extracted proteins and peptides were purified by precipitation and then fluorescently labeled with Cy2, Cy3, and Cy5 fluorophores. The Cy2 dye was intended to label the internal standard, which was a sample containing an equilibrium mixture of proteins from all analyzed samples. Individual samples were cyclically labeled with Cy3 and Cy5 dyes. Labeled samples of proteins and peptides were analyzed by 2D-DIGE electrophoresis (two-dimensional differential gel electrophoresis) [88]. For this purpose, isoelectric focusing of the peptides was performed on a polyacrylamide gel in a pH 3–10 gradient (first dimension—IEF), and then the focused peptides were

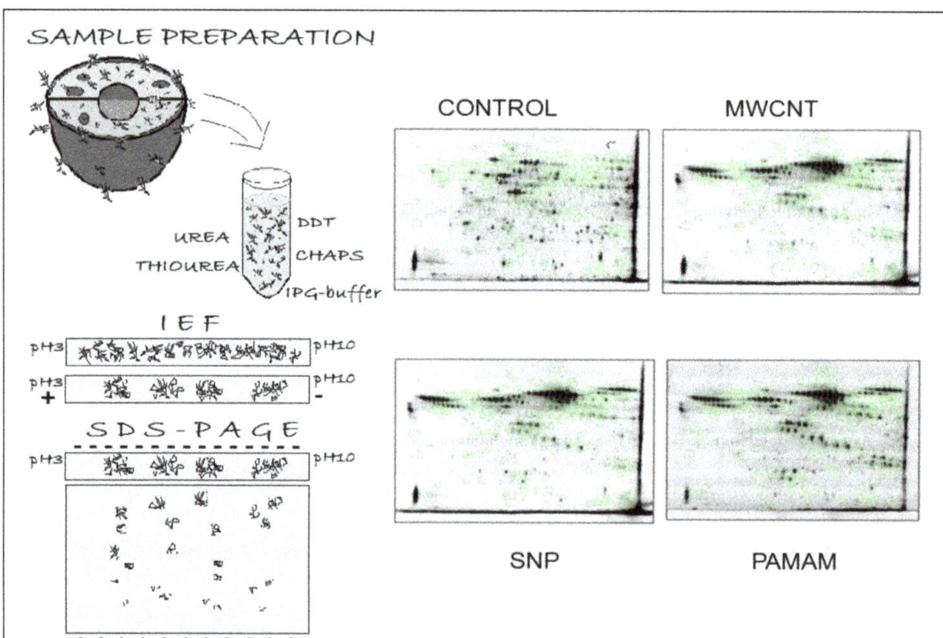

Figure 4.4. Schematic presentation of an experiment to assess changes in the peptide (proteome) profile of cells in response to stress (contact with multi-walled carbon nanotubes (MWCNTs), nanosilver particles (SNPs), and PAMAM dendrimers). Explanations in the text below. Green boundaries show the result of the detection of spots on the gel.

separated in the second dimension in a standard 12.5% polyacrylamide gel. After the electrophoretic separation was fully developed, fluorescence scans of the gels were made with the use of a three-color fluorescence scanner.

The fluorescence technique allowed for obtaining separate images of gels for individual fluorescent markers and assigning them to appropriate samples of nanoparticles. An exemplary result of the experiment is shown in figure 4.4. As can be seen, all the nanomaterials used were the cause of the changes in the pattern of spots visible on the gels in relation to the gel obtained for control cells. Moreover, significant differences can also be seen in the proteome profile between individual cell samples when exposed to different nanomaterials, although not as extensive as compared to the control. Therefore, we conclude that there are large and specific changes in the peptide profile (proteome) of cells exposed to various nanomaterials. Quantitative analysis of differences in the protein profile allows the identification of spots with a significant difference in expression. With the use of mass spectrometry dedicated to proteomics [89], it is possible to identify individual peptides and proteins with altered expression in cells exposed to the tested nanomaterials.

4.7 Biological recognition of stress

A number of our own observations made with transcriptomics and proteomics methods and the obtained data confirm our belief that the metabolic response of cells in the form of changes in gene expression and changes in the peptide and protein profile of the cell may be useful for the identification of materials engineering and nanotechnology products. It found its reflection in the patent granted to us [90]. This patent concerns the use of cells as a specific sensor recognizing the surface of the material or nanomaterial particle with which the cell is in contact, and the recognition process itself can be described as the biological recognition of the stress factor, which is this contact.

When analyzing the metabolic response of a cell exposed to the above stress factors, it was observed that certain groups of genes change in expression in a similar way (in terms of the direction of the changes) regardless of the type of surface used or the nanoparticles used, while other genes are altered in a highly specific manner characteristic of a given stress factor. The first group is characterized by a non-specific metabolic response, common to all stress factors in relation to a specific cell and specific culture conditions, while the remaining genes carry information about a specific response.

Since not all changes in gene expression (transcriptome) translate into changes in the protein and peptide composition of cells (proteome), more reliable information about the metabolic response can be obtained by analyzing both the transcriptome profile and the proteome profile. The use of this strategy resulted in the possibility of recognizing and distinguishing the surfaces of two titanium alloys (Ti6Al4V and Ti6Al7Nb), additionally subjected to two types of surface modification, and the polished surface of 316L medical steel. The obtained transcriptome and proteome profiles for these samples were saved (as response patterns). In the next experiment, the above-mentioned samples and similarly prepared two additional steel samples with different chemical compositions were used. All samples, with identical shapes and dimensions, were coded and used as stress factors just as in the previous experiment. After obtaining transcriptome and proteome profiles for each of the samples, the obtained profiles were compared with the response patterns from the first experiment. As a result, the obtained profiles were in accordance with the response patterns, and this consistency was within the range of 85%–90% for the transcriptome profile and within the range of 65%–80% for the proteome profile of the identified samples. For samples of unknown chemical composition, the similarity of the obtained transcriptome and proteome profiles with the predefined patterns was significantly lower and amounted to approximately 35%–38% for the transcriptome and 45%–50% for the proteome.

In a separate, similar experiment, bacterial cells were used in contact with the surface of polished 316L medical steel covered with a layer of nano-crystalline diamond and with diamond powders of various grain sizes, administered as a suspension under forced laminar flow conditions. For this experiment, only the transcriptome profiles were analyzed and appropriate patterns were prepared. Then the same bacterial cell suspension was used to identify the same (coded) samples and additionally a sample of unknown material applied in the form of a powder. Also in this case, a very high compliance of the transcriptome profiles of the tested materials with the standards response patterns was obtained (over 90%), while in the case of an unknown powder sample, the similarity of the obtained profile with the standards did not exceed 60%–70%.

The specificity of the metabolic response of cells has also been confirmed in other our experiments in response to stress resulting from exposure of cells to contact with biomaterials [91–93] and nanoparticles [94]. The results discussed above show that it is possible to distinguish not only materials with different chemical compositions, but also allow to differentiate the type of surface development for materials with the same chemical composition, and in the case of nanoparticles, it is also possible to differentiate their sizes. The proposed strategy could be used in analogy to the existing databases used to identify chemical compounds by mass spectrometry. However, this strategy requires first gathering appropriate metabolic response patterns obtained for defined cells under strictly defined culture conditions and storing these patterns in an available database. It can be assumed that the cBiT database prepared for the transcriptome already exists, only to fill it with reliable research results.

In summary, biological recognition resulting from molecular recognition, being an integral part of materiomics, provides a useful tool for the detection and identification of materials and nanomaterials. Thus, materiomics, which is a platform linking the materials science with the life sciences, constitutes a potential for development and impact on both disciplines that cannot be overestimated. The near future will show this potential in all its glory.

References

[1] The Book of Genesis *The Vatican* https://vatican.va/archive/bible/genesis/documents/bible_genesis_en.html

[2] Montgomery D R 2012 The evolution of creationism *GSA Today* **22** 4–9

[3] Darwin C 2011 *On the Origin of Species: A digital edition of 1859 London Origin of Species* https://darwin.amnh.org/files/images/pdfs/e83461.pdf

[4] Gonzalez W J (ed) 2008 *Evolutionism: Present Approaches* (A Coruña: Netbiblio)

[5] Noble A 2010 *An Introduction to Intelligent Design* (Glasgow: Centre for Intelligent Design) https://ripeideas.co.uk/wp-content/uploads/2016/01/C4iD-Book-One.pdf

[6] Petkowski J J, Bains W and Seager S 2020 On the potential of silicon as a building block for life *Life* **10** 84

[7] Sun K 2016 Complex organics in space from solar system to distant galaxies *Astron. Astrophys. Rev.* **24** 8

[8] Vakoch D A 2013 Life beyond Earth and the evolutionary synthesis *Astrobiology, History, and Society, Advances in Astrobiology and Biogeophysics* ed D A Vakoch (Berlin: Springer) ch 9

[9] Hub Z 2020 Friedrich Engels and the technoscientific reproducibility of life: synthetic cells as case material for practicing dialectics of science today *Sci. Soc.* **84** 369–400

[10] Tekstowo.pl *Skaldowie - Nie ma szatana* https://tekstowo.pl/piosenka,skaldowie,nie_ma_szatana.html

[11] Fruton J S and Simmonds S 1958 *General Biochemistry* 2nd edn (New York: Wiley)

[12] Rodwell V, Bender D, Weil P A, Kennelley P and Botham K 2018 *Harper's Illustrated Biochemistry* 30th edn (New York: McGraw-Hill)

[13] Crick F H C 1958 On protein synthesis *Symp. Soc. Exp. Biol.* **12** 138–63 https://wellcomecollection.org/works/z3d5fnyg
Crick F H C 1970 Central dogma of molecular biology *Nature* **227** 561–3

[14] Watson J D 1976 *Molecular Biology of the Gene* ed W A Benjamin (Menla Park, CA: Harvard University and Cold Spring Harbor Laboratory)

[15] Coffin J M and Fan H 2016 The discovery of reverse transcriptase *Annu. Rev. Virol.* **3** 29–51

[16] Bussard A E 2005 A scientific revolution? *EMBO Rep.* **6** 691–4

[17] Prusiner S B 1982 Novel proteinaceous infectious particles cause scrapie *Science* **216** 136–44

[18] Manix M, Kalakoti P, Henry M, Thakur J, Menger R, Guthikonda B and Nanda A 2015 Creutzfeldt–Jakob disease: updated diagnostic criteria, treatment algorithm, and the utility of brain biopsy *Neurosurg. Focus* **39** E2

[19] Bastian F O, Sanders D E, Forbes W A, Hagius S D, Walker J V, Henk W G, Enright F M and Elzer P H 2007 *Spiroplasma spp.* from transmissible spongiform encephalopathy brains or ticks induce spongiform encephalopathy in ruminants *J. Med. Microbiol.* **56** 1235–42

[20] Bastian F O 2014 The case for involvement of spiroplasma in the pathogenesis of transmissible spongiform encephalopathies *J. Neuropathol. Exp. Neurol.* **73** 104–14

[21] Fraser C M *et al* 1995 The minimal gene complement of *Mycoplasma genitalium Science* **270** 397–403

[22] Glass J I, Assad-Garcia N, Alperovich N, Yooseph S, Lewis M R, Maruf M, Hutchison C A III, Smith H O and Venter J C 2006 Essential genes of a minimal bacterium *Proc. Natl Acad. Sci.* **103** 425–30

[23] Gibson D G *et al* 2010 Creation of a bacterial cell controlled by a chemically synthesized genome *Science* **329** 52–6

[24] Callaway E 2016 Minimal cell raises stakes in race to harness synthetic life *Nature* **531** 557–8

[25] Mikołaj Śmigiel W, Lefrançois P and Poolman B 2019 Physicochemical considerations for bottom-up synthetic biology *Emerg. Top. Life Sci.* **3** 445–58

[26] Abil Z and Danelon C 2020 Roadmap to building a cell: an evolutionary approach *Front. Bioeng. Biotechnol.* **8** 927

[27] Elani Y 2021 Interfacing living and synthetic cells as an emerging frontier in synthetic biology *Angew. Chem.* **133** 5662–71

[28] Fanalista F, Birnie A, Maan R, Burla F, Charles K, Grzegorz Pawlik, Deshpande S, Koenderink G H, Dogterom M and CeesDekker 2019 Shape and size control of Artificial cells for bottom-up Biology *ACS Nano* **13** 5439–50

[29] Exterkate M and Driessen A J M 2019 Synthetic minimal cell: self-reproduction of the boundary layer *ACS Omega* **4** 5293–303

[30] Birnie A and Dekker C 2021 Genome-in-a-box: building a chromosome from the bottom up *ACS Nano* **15** 111–24

[31] Vibhute M A, Schaap M H, Maas R J M, Nelissen F H T, Spruijt E, Heus H A, Hansen M M K and Huck W T S 2020 Transcription and translation in cytomimetic protocells perform most efficiently at distinct macromolecular crowding conditions *ACS Synth. Biol.* **9** 2797–807

[32] Pelletier J F *et al* 2021 Genetic requirements for cell division in a genomically minimal cell *Cell* **184** 2430–40

[33] Dovidchenko N V, Leonova E I and Galzitskaya O V 2014 Mechanisms of amyloid fibril formation *Biochemistry (Moscow)* **79** 1515–27

[34] Hartl F U, Bracher A and Hayer-Hartl M 2011 Molecular chaperones in protein folding and proteostasis *Nature* **475** 324–32

[35] Bhatwa A, Wang W, Hassan Y I, Abraham N, Li X-Z and Zhou T 2021 Challenges associated with the formation of recombinant protein inclusion bodies in *Escherichia coli* and strategies to address them for industrial applications *Front. Bioeng. Biotechnol.* **9** 630551

[36] Schneider H-J 2009 Binding mechanisms in supramolecular complexes *Angew. Chem. Int. Ed.* **48** 3924–77

[37] Houk K N, Leach A G, Kim S P and Zhang X 2003 Binding affinities of host–guest, protein–ligand, and protein–transition-state complexes *Angew. Chem. Int. Ed.* **42** 4872–97

[38] Thompson C J and McBride J L 1974 On Eigen's theory of the self-organization of matter and the evolution of biological macromolecules *Math. Biosci.* **21** 127–42

[39] Percec V, Xiao Q, Lligadas G and Monteiro M J 2020 Perfecting self-organization of covalent and supramolecular mega macromolecules via sequence-defined and monodisperse components *Polymer* **211** 123252

[40] Baron R and McCammon J A 2013 Molecular recognition and ligand association *Annu. Rev. Phys. Chem.* **64** 151–75

[41] Martins J N, Carlos Lima J and Basílio N 2021 Selective recognition of amino acids and peptides by small supramolecular receptors *Molecules* **26** 106

[42] Alberti S 1997 The origin of the genetic code and protein synthesis *J. Mol. Evol.* **45** 352–8

[43] Meisenberg G and Simmons W H 2012 DNA, RNA, and protein synthesisPrinciples of Medical Biochemistry (Amsterdam: Elsevier) ch 9 pp 66–93

[44] Irfan-ur-Rauf T, Ali F, Dar J S, Magray A R, Ganai B A and Chishti M Z 2019 Posttranslational modifications of proteins and their role in biological processes and associated diseases *Protein Modificomics* (Amsterdam: Elsevier) ch 1 pp 1–35

[45] Kurdyukov S and Bullock M 2016 DNA methylation analysis: choosing the right method *Biology* **5** 3

[46] Squire J 2019 The actin–myosin interaction in muscle: background and overview *Int. J. Mol. Sci.* **20** 5715

[47] Ananthakrishnan R and Ehrlicher A 2007 The forces behind cell movement *Int. J. Biol. Sci.* **3** 303–17
[48] Ballester P and Vidal-Ferran A 2008 Introduction to supramolecular catalysis *Supramolecular Catalysis* ed P W N M van Leeuwen (Weinheim: Wiley)
[49] Miyamura H, Bergman R G, Raymond K N and Toste F D 2020 Heterogeneous supramolecular catalysis through immobilization of anionic M4L6 assemblies on cationic polymers *J. Am. Chem. Soc.* **142** 19327–38
[50] Lim N C H and Jackson S E 2015 Molecular knots in biology and chemistry *J. Phys. Condens. Matter* **27** 354101
[51] Zhao L, Cai J, Li Y, Wei J and Duan C 2020 A host–guest approach to combining enzymatic and artificial catalysis for catalyzing biomimetic monooxygenation *Nat. Commun.* **11** 2903
[52] Parrot A, Collin S, Bruylants G and Reinaud O 2018 The 3rd degree of biomimetism: associating the cavity effect, Zn^{II} coordination and internal base assistance for guest binding and activation *Chem. Sci.* **9** 5479
[53] Dong Z, Luo Q and Liu J 2012 Artificial enzymes based on supramolecular scaffolds *Chem. Soc. Rev.* **41** 7890–908
[54] Petsko G A and Yates J R III 2011 Analyzing molecular interactions *Curr. Protoc. Bioinform.* **36** 8.1.1–4
[55] Nair A, Chauhan P, Saha B and Kubatzky K F 2019 Conceptual evolution of cell signaling *Int. J. Mol. Sci.* **20** 3292
[56] Uings I J and Farrow S N 2000 Cell receptors and cell signalling *J. Clin. Pathol.: Mol. Pathol.* **53** 295–9
[57] Rye C, Wise R, Jurukovski V, DeSaix J, Choi J and Avissar Y 2016 Cell communication *Biology* (Houston, TX: OpenStax) ch 9 https://openstax.org/books/biology/pages/9-1-signaling-molecules-and-cellular-receptors
[58] Chaplin D D 2006 Overview of the human immune response *Am. Acad. Allerg. Asthma Immunol.* **117** S430–5
[59] 2007 Understanding the Immune System—How It Works *NIH Publication* No. 07-5423 National Institute of Allergy and Infectious Diseases http://imgt.org/IMGTeducation/Tutorials/ImmuneSystem/UK/the_immune_system.pdf
[60] Hiam-Galvez K J, Allen B M and Spitzer M H 2021 Systemic immunity in cancer *Nat. Rev.* **21** 345–9
[61] Wucherpfenniga K W and Sethi D 2011 T cell receptor recognition of self and foreign antigens in the induction of autoimmunity *Semin. Immunol.* **23** 84–91
[62] Orensteina W A and Ahmed R 2017 Simply put: vaccination saves lives *Proc. Natl Acad. Sci.* **114** 4031–3
[63] Noh J Y, Jeong H W and Shin E-C 2021 SARS-CoV-2 mutations, vaccines, and immunity: implication of variants of concern *Signal Transduct. Target. Ther.* **6** 203
[64] Perkins J R, Diboun I, Dessailly B H, Lees J G and Orengo C 2010 Transient protein–protein interactions: structural, functional, and network properties *Structure* **18** 1233–43
[65] Baron R and McCammon J A 2013 Molecular recognition and ligand association *Annu. Rev. Phys. Chem.* **64** 151–75
[66] Chambers S S *et al* 2000 A review of molecular recognition technologies for detection of biological threat agents *Biosens. Bioelectron.* **15** 549–78

[67] Magdeldin S and Moser A 2012 *Affinity Chromatography: Principles and Applications* ed S Magdeldin (Rijeka: InTech)

[68] Alhabbab R Y 2018 *Basic Serological Testing* (Berlin: Springer)

[69] Englebienne P, Van Hoonacker A and Verhas M 2003 Surface plasmon resonance: principles, methods and applications in biomedical sciences *Spectroscopy* **17** 255–73

[70] Velazquez-Campoy A, Ohtaka H, Nezami A, Muzammil S and Freire E 2004 Isothermal titration calorimetry *Curr. Protoc. Cell Biol.* **23** 17.8.1–24

[71] Hajighasemi F, Saboor-Yaraghi A A and Shokr F 2004 Measurement of affinity constant of anti-human IgG monoclonal antibodies by an ELISA-based method *Iran J. Immunol.* **1** 154–61

[72] Leavitt S and Freire E 2001 Direct measurement of protein binding energetics by isothermal titration calorimetry *Curr. Opin. Struct. Biol.* **11** 560–6

[73] Crescenzo G D, Boucher C, Durocher Y and Jolicoeur M 2008 Kinetic characterization by surface plasmon resonance-based biosensors: principle and emerging trends *Cell. Mol. Bioeng.* **1** 204–15

[74] Cobb J P, Hotchkiss R S, Karl I E and Buchman T G 1996 Mechanisms of cell injury and death *Br. J. Anaesth.* **77** 3–10

[75] Lee P, Chandel N S and Simon M C 2020 Cellular adaptation to hypoxia through HIFs and beyond *Nat. Rev. Mol. Cell Biol.* **21** 268–83

[76] Wellen K E and Thompson C B 2010 Cellular metabolic stress: considering how cells respond to nutrient excess *Mol. Cell* **40** 323–32

[77] Żur J, Wojcieszyńska D and Guzik U 2016 Metabolic responses of bacterial cells to immobilization *Molecules* **21** 958

[78] van der Knaap J A and Verrijzer C P 2016 Undercover: gene control by metabolites and metabolic enzymes *Genes Develop.* **30** 2345–69

[79] Towle H C 1997 Metabolic regulation of gene expression *Emerging Technologies for Nutrition Research* ed S J Carlson-Newberry and R B Costello (Washington, DC: National Academy Press) pp 375–88 ch 17

[80] Castagnola V, Zhao W, Boselli L, Lo Giudice M C, Meder F, Polo E, Paton K R, Backes C, Coleman J N and Dawson K A 2018 Biological recognition of graphene nanoflakes *Nat. Commun.* **9** 1577

[81] Walczynska M, Jakubowski W, Wasiak T, Kadziola K, Bartoszek N, Kotarba S, Siatkowska M, Komorowski P and Walkowiak B 2018 Toxicity of silver nanoparticles, multiwalled carbon nanotubes, and dendrimers assessed with multicellular organism *Caenorhabditis elegans Toxicol. Mech. Meth.* **28** 432–9

[82] Kim J H, Lee S H, JeongCha Y, Hong S J, Chung S K, Park T H and Choi S S 2016 *C. elegans*-on-a-chip for *in situ* and *in vivo* Ag nanoparticles' uptake and toxicity assay *Sci. Rep.* **7** 40225

[83] Siatkowska M *et al* 2021 Impact of micron-sized diamond particles on barrier cells of the human small intestine *Diam. Relat. Mater.* **114** 108307

[84] Sokołowska P, Białkowska K, Siatkowska M, Rosowski M, Kucinska M, Komorowski P, Makowski K and Walkowiak B 2017 Human brain endothelial barrier cells are distinctly less vulnerable to silver nanoparticles toxicity than human blood vessel cells: a cell-specific mechanism of the brain barrier? *Nanomed.-Nanotechnol. Biol. Med.* **13** 2127–30

[85] Kolodziejczyk A M, Sokołowska P, Zimon A, Grala M, Rosowski M, Siatkowska M, Komorowski P and Walkowiak B 2021 Dysfunction of endothelial cells exposed to nanomaterials assessed by atomic force spectroscopy *Micron* **145** 103062

[86] Roche Diagnostics 2006 *PCR Applications Manual* 3rd edn (Mannheim: Roche Diagnostics) https://gene-quantification.de/ras-pcr-application-manual-3rd-ed.pdf
[87] Khelurkar V C, Ingle K P and Padole D 2017 DNA microarray: basic principle and its applications *Trends Biosci.* **10** 488–90
[88] GE Healthcare 2005 *Ettan DIGE System User Manual* 18-1173-17 Edition AB GE Healthcare https://mcgill.ca/cian/files/cian/ge_dige_manual.pdf
[89] Bantscheff M, Schirle M, Sweetman G, Rick J and Kuster B 2007 Quantitative mass spectrometry in proteomics: a critical review *Anal. Bioanal. Chem.* **389** 1017–31
[90] Walkowiak B, Komorowski P and Walkowiak-Przybyło M 2016 Means of identification of materials engineering products or nanotechnologies (in Polish: Sposób identyfikacji produktów inżynierii materiałowej lub nanotechnologii) *Polish Patent* PL 223224 granted on 31.10
[91] Komorowski P, Walkowiak-Przybyło M and Walkowiak B 2015 Early cell response to contact with biomaterial's surface *J. Biomed. Mater. Res.* B **104** 880–93
[92] Walkowiak-Przybyło M, Komorowski P and Walkowiak B 2017 Differences in the expression of cell cycle genes in osteoblasts and endothelial cells cultured on the surfaces of Ti6Al4V and Ti6Al7Nb alloys *J. Biomed. Mater. Res. Part* A **105** 1607–17
[93] Komorowski P *et al* 2020 Comprehensive biological evaluation of biomaterials used in spinal and orthopedic surgery *Materials* **13** 4769
[94] Komorowski P *et al* 2019 Simultaneous transcriptome and proteome analysis of EA.hy926 cells under stress conditions induced by nanomaterials *J. Biomed. Mater. Res.* B **107** 1024–034

Chapter 5

Quantum basis of materials recognition by cells

An innovative approach for the identification of nanomaterials is proposed and discussed, based on two main observations. First, as the dimensions of objects decrease, it is observed that quantum effects play a greater role. Second, it was found that the metabolic response of cells to stress resulting from contact with the surfaces of nanostructures or materials is of a very specific nature. In other words, the cell is able to recognize the source of the stress and can serve as a specific sensor that recognizes the nanostructure and surface of the material. Postulates and arguments supporting this strategy are presented, and at the end some practical comments resulting from own experience are presented.

5.1 A new strategy for the identification and grouping of nanomaterials

Progress in the field of innovative technologies, in particular nanotechnology, is an invaluable source of ideas to create new materials with previously unattainable properties and possible applications in all areas of human activity. Unfortunately, the products of nanotechnology are not only the basis for further intensive technological progress, but may also contribute to an as yet unidentified risk to our environment, including health threats to living organisms including humans. Hence there is an urgent need to identify and control these risks. The authorities of different countries take action in this regard. For example, in the European Union, activities in this area are institutionally subject to the European Chemicals Agency, which is trying, in cooperation with other European agencies and competent representatives of the member states, to implement rules for the registration of nanomaterials. The process of risk assessment is also in the focus of researchers. For example within the EU FP7 project concerning Managing the Risks of Nanoparticles (MARINA), a strategy for grouping and read-across approaches has been developed [1, 2], but the interest in the problem and attempts to solve it are global, which is very well illustrated in the review of available literature in [3].

doi:10.1088/978-0-7503-2656-8ch5

The main problem in this regard is the poor capability to identify and group the engineered products of nanotechnology on the basis of standard physicochemical methods, which work perfectly at the macro-scale. It turns out that the same materials at the nano-scale exhibit extremely different physicochemical properties and biological reactivity depending on the size of the particles. The situation is further complicated by the treatment of the surface of nanoparticles in order to stabilize them or to make intentional changes in their chemical and biological reactivity. Despite the use of very advanced analytical techniques, so far the development of a standard procedure for identifying, classifying, and grouping the products of nanotechnology has failed.

The innovative approach proposed and discussed below is based on two main observations. First, with the decreasing dimensions of objects, the more significant are the quantum effects, which are responsible for the unusual properties of nanostructures [4]. Second, it was found that the metabolic response of cells to stress resulting from contact with the surfaces of nanostructures or materials is very specific in its nature. In other words, the cell is able to recognize a source of stress and can be used as a specific sensor which recognizes a nanostructure [5, 6] and a material surface [7, 8]. The same is true for simple multicellular organisms, such as the nematode *Caenorhabditis elegans* [9, 10]. This should not come as a particular surprise, because cellular biochemical processes are carried out at the level of interactions of nanostructures (the biomolecules involved in this process are also specialized nanostructures).

Both observations correspond well to the strategy called materiomics discussed in previous chapters, which includes systematic studies of the relationships between composition, processing, structure, properties, and performance of advanced materials. So far materiomics has particular applications for biological materials and structures [11] and biomaterials [12, 13].

These two observations underlie the proposed intelligent strategy that is useful for the classification and grouping of nanomaterials based on the analysis of the similarities and differences in the measured electron orbitals and established profiles in the expression of genes and peptides. Together with the so far accumulated results of the classical analysis of physicochemical and eco-toxicological properties, this strategy will allow for the effective implementation of nano-QSAR models [14] and read-across strategies [1, 15] for classifying and grouping engineered nanomaterials.

5.2 Postulates

Considering the above and combining it with the previously discussed issues of biological recognition, the following postulates have been formulated:

1. Proper characterization, classification, and grouping of nanomaterials require taking into account the quantum effects of electrons present on their surface. Both electron orbitals and their self-energy levels can be determined using scanning tunneling microscopy (STM) and photoelectron spectroscopy (PES) techniques, respectively, and similarities and differences in the measured orbitals and self-energies can be used in the identification, classification, and grouping of nanomaterials.

2. The metabolic response of cells to stress resulting from the contact with nanostructures is unique and very specific in nature due to the quantum mechanisms responsible for the creation, by an interaction with nanostructure, a new bonding orbital or breaking of current bonds by induction of antibonding orbitals in the receptor–effector complex and start a specific pathway of cellular response. This molecular response of cells can be observed using transcriptomics and proteomics methods and can be used for the identification, classification, and grouping of nanomaterials.
3. Both approaches are complementary.

5.3 Supporting arguments

In contrast to the commonly used physicochemical methods, our concept relies on two complementary approaches: the estimation and interpretation of the quantum properties of nanomaterials, and the estimation and interpretation of the biological processes of the recognition of nanomaterials by cells.

5.4 The quantum effects of nanomaterials

The properties of nanomaterials have been discussed extensively in chapter 2, and particularly in the section 'Problems with nanomaterials'. Therefore, here we only present selected arguments briefly. The unexpected properties of nanomaterials are mainly a result of geometrical effects. This means that with a decrease in a linear dimension of the object (for example diameter), the surface to volume ratio (S/V) increases. Starting from 10^{-7} m this increase is very rapid and significant. It also means that the percentage of atoms present on the surface also increases up to the critical point, where all atoms are available on the surface.

The geometrical effect (an increase of the S/V) is accompanied by the manifestation of quantum effects. For nanoparticles with a diameter of 100 nm these effects are rather weak, but decreasing the particle size quantum effects become more and more significant. For example, when suspended in water, gold nanoparticles produce different colors of the suspension depending on the particles' size [4, 16]. Also, the energy levels (states) on the surface of nanoparticles change in comparison to the bulk substance [4]. Well known and useful in practice is the surface plasmon resonance (SPR), which is also a quantum effect [17]. The field emission of electrons from a sharp spike can be realized with low voltage applied in a nano-spike geometry, whereas in a macro-spike geometry it requires kilovolts for a comparable current density [18]. Mechanical properties, such as an increase in hardness and elastic modulus [19], as well as a decrease in melting point [20], are also significantly affected in the nano-scale.

The geometrical effect is also essential in biological systems. For example, the electric field present across the cell membrane (thickness about 10 nm) in the resting state (resting voltage potential is about 100 mV) is very strong and reaches 10^7 V m^{-1}. Such a high value is rare and very dangerous at the macro-scale. This strong electric field is able to keep the membrane (a kind of capacitor) in a state of high rigidity, but during cell activation and membrane depolarization the membrane becomes much more fluid, allowing for substances to flow through the membrane.

The characterization and grouping of substances was one of the main activities in research at the end of the nineteenth century. Many researchers tried to find order in nature. The first person who reached success in substance grouping was Mendeleev, and his table of periodicity of chemical elements remains in force to date. Mendeleev based his table on the physicochemical properties of chemical substances, mainly on substance density and chemical reactivity. This approach is good enough for the characterization and grouping of pure chemical substances, but it is not sufficient for complicated nanoparticles, even with the use of contemporary available physical and chemical experimental methods. This is the basic and so far unsolved problem in the discussion of the grouping of nanomaterials and read-across procedures.

Let us look for a moment into quantum physics, where the Schrödinger equation allows us to find the *self-energy* E_n of the state described by the *wave function* ψ_n, permitted for an electron in the field of *energy potential* V. The wave function determines the area of space where the probability of finding an electron is high. This means the wave function determines the orbital with the permitted state of energy (figure 5.1). In light of the above, contemporary researchers would have no problem with grouping chemical elements according to orbitals and their energy states resulting from Schrödinger's equation. Thus the Mendeleev's table of elements would look like the table of orbitals, available as a poster in [21]. So the question arises: why are nanomaterials not grouped according to orbitals and their energies?

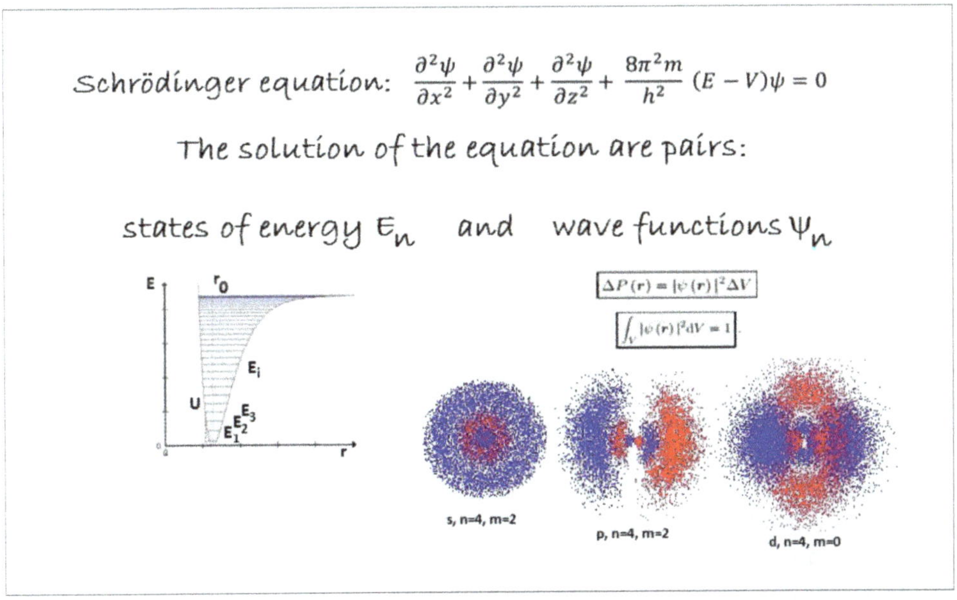

Figure 5.1. The Schrödinger equation allows us to find the self-energy E_n of the state described by the wave function ψ_n, permitted for an electron in the field of energy potential V. The wave function determines the area of space where the probability of finding an electron is high. This means the wave function determines the orbital with the permitted state of energy. The s, p, and d orbitals of the silver atom were visualized with the Orbital Viewer software made available by David Manthey [22].

Let us use this way of thinking for the analysis of nano objects. Starting from a macroscopic substance, the electrons present at the surface feel the energy potential V derived from all the atoms of this particle, which is rather constant over the whole surface because of the enormous number of atoms that together create this potential. Splitting the substance to smaller pieces the same potential V is present for a long time, until the size of the particle reaches the nano-scale (below 100 nm) and the number of atoms making up the nanostructure becomes greatly reduced. At this point variation in the value of energy potential V also causes changes of the available electron orbitals and its self-energies, resulting from solutions of the Schrödinger equation. Any change of the energy potential V is accompanied by a change of the available orbitals' shape and self-energies, and also by the nanoparticle's physicochemical properties. The same conclusions will be obtained starting from a singular atom and increasing the size and complexity of the nanoparticle. At the beginning, the addition of a single atom results in a significant change in the value of the V potential, and thus the self-states of energy and wave functions also change. As more atoms are added, these changes become less and less important, and finally the value of the energy potential V becomes constant, which means the transition to the macro-scale.

We have tools available for this kind of analysis. For theoretical considerations, quantum physics and density functional theory (DFT) can be applied. Experimental approaches are also available. The orbital's shape can be visualized with scanning tunneling microscopy (STM). Examples of theoretical considerations and experimental results for the visualization of orbitals can be found in the experimental paper [23]. States of electron energy corresponding to the indicated orbitals can also be measured. Commercially available tools enabling orbital energy analysis are photoelectron spectroscopy (PES) systems. As an example of orbital energy measurement, and the resulting space presentation of orbitals maps, the orbital tomography approach can be considered [24].

At this point, we again express our wonderment at the consistent use of methods, admittedly well proven on a macroscopic scale, to characterize and group nanomaterials. These methods do not necessarily work well at the nano-scale. Why are people stubbornly trying to group nanomaterials on the basis of their physicochemical properties, including, for example, size, surface roughness, and its modification, and other parameters well tested on a macroscopic scale, instead of taking into account orbitals and their corresponding self-energies? Why is it not surprising that the properties of various chemical compounds change depending on their atomic composition and the bonds between atoms, and it is expected that in the case of nanoparticles and nanostructures this effect will be insignificant? Let us try, for example, to consider the variability of the properties of the simplest carbon and hydrogen compounds in the form of alkanes, i.e. saturated hydrocarbons. By starting from the simplest compound consisting of one carbon atom and four hydrogen atoms (CH_4—methane) and adding further carbon and hydrogen atoms, an increasingly longer carbon chain is obtained, the chemical composition of which is described by the relationship C_nH_{2n+2}. The ratio of the number of hydrogen atoms to the number of carbon atoms is $2(1+1/n)$ and as the number n of carbon atoms increases, it tends to the value of 2. Each addition of one carbon and two hydrogen atoms lengthens the chain, practically without changing its transverse dimension, and significantly changes its physicochemical properties. Since the single C–C and C–H covalent bonds are nonpolar, a large amount of energy is required to break them. For this reason, the chemical reactivity of alkanes is limited to the combustion or substitution of a hydrogen atom, most often with a chlorine atom. Pyrolysis leading to a reduction in the chain length is also possible. As an example, consider the first ten alkanes, starting with methane and ending with decane. Figure 5.2 shows the values of the melting, boiling, and auto-ignition temperatures, as well as the surface tension, the refractive index, and the dielectric constant. As can be seen, these parameters change significantly as a function of the number of carbon atoms in the molecule.

Similar observations are made for linear unsaturated hydrocarbons, including alkenes (C_nH_{2n}), containing one double bond between carbon atoms, and alkynes (C_nH_{2n-2}), having one triple bond. For alkenes, the ratio of the number of hydrogen atoms to the number of carbon atoms is 2, while for alkynes it tends to 2 as the chain length increases (figure 5.3), but the properties of alkanes, alkenes, and alkynes do not become the same. Alkenes are more chemically reactive than alkanes and, apart from combustion and substitution reactions, also undergo polymerization reactions and the addition of chlorine, bromine, or water. Alkynes are even more reactive than alkanes and alkenes, which makes them unstable and subject to many spontaneous reactions.

This class of hydrocarbons is further enriched by their branched forms in which at least one carbon atom is bonded to at least three other carbon atoms but does not form cyclic forms. Structural differences strongly influence the properties of hydrocarbons. The information presented here belongs to the basics of organic chemistry and does not contain any elements of novelty, but also draws our attention to the fact that the properties of these chemical compounds are decisively influenced by the bonds between carbon and hydrogen as well as carbon and carbon atoms. Here, too, we can use information from the basics of organic chemistry. These bonds are formed by the sharing of electrons between atoms, or more precisely, by mixing the electronic orbitals of individual atoms and the formation of bonding molecular orbitals.

Mixing atomic orbitals can result in the formation of sp, sp^2, and sp^3 molecular hybrids. In the case of alkanes we are talking about sp^3 hybridization. In the case of alkenes, the double bond between two carbon atoms is realized by one sigma σ bond resulting from the overlapping of the frontally interacting sp2 orbitals and one pi (π) bond resulting from the overlapping of the side non-hybridized p orbitals, while in the case of alkynes, the triple bond between carbon atoms is realized by one σ bond and two π bonds. The strongest bond and that with the shortest distance between two carbon atoms is the triple bond, the double bond takes an intermediate position, and the single bond is the weakest and has the greatest distance between atoms. More information on hybrids of atomic orbitals in the form of molecular orbitals and the formation of bonds in the molecule can be found in any introductory textbook on organic chemistry [25].

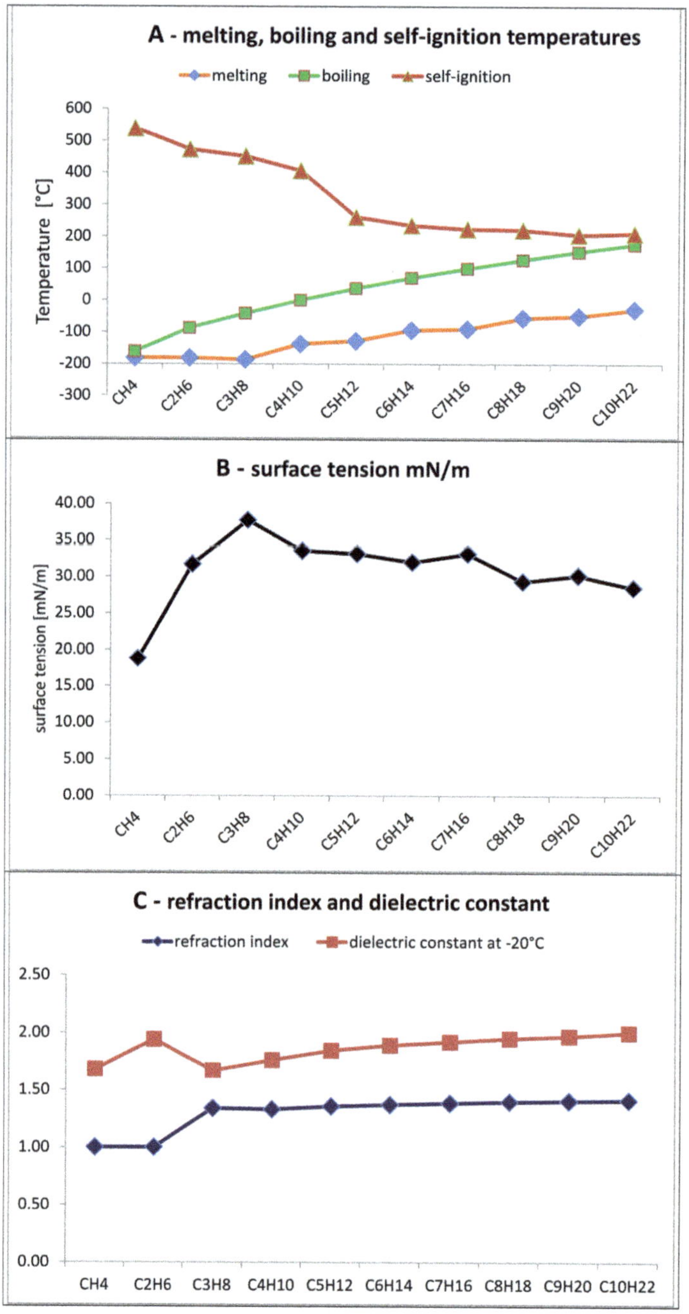

Figure 5.2. Comparison of the values of selected physicochemical parameters of alkanes, starting with methane and ending with decane. A—melting, boiling, and auto-ignition temperatures; B—surface tension; C—refractive index and relative dielectric constant. The charts were prepared using data generally available on the Internet.

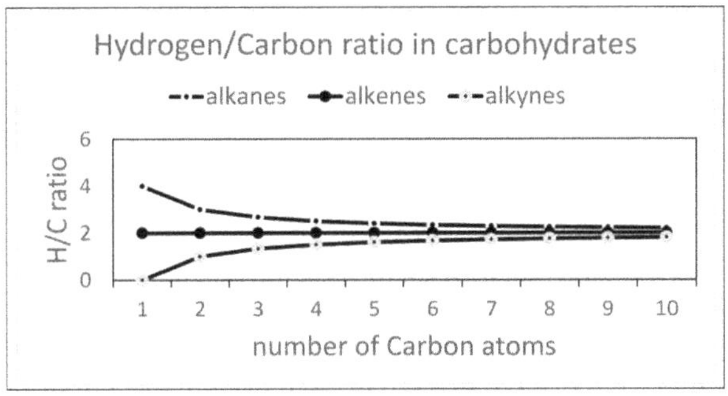

Figure 5.3. The ratio of the number of hydrogen atoms to the number of carbon atoms in the chains of alkanes, alkenes, and alkynes.

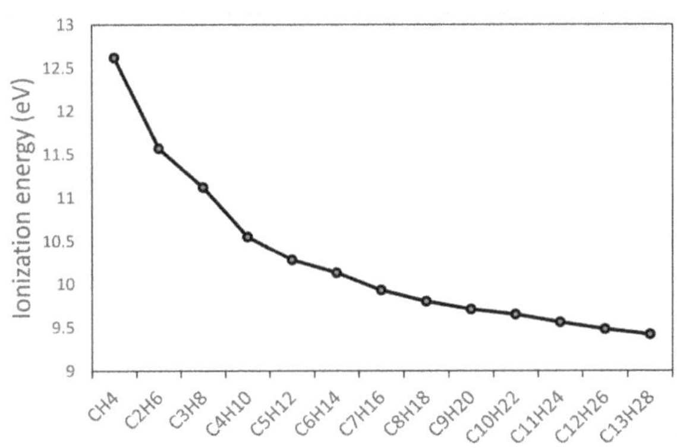

Figure 5.4. The dependence of the first ionization energy on the number of carbon atoms in the alkane molecule. Based on data from [26, 27].

It is also worth noting that not only molecular orbitals are characteristic of a given molecule, but also the ionization energies of the molecule, depending on the bonds present, are a feature that characterizes the molecule. Figure 5.4 shows the dependence of the first ionization energy (i.e. the energy necessary for a single ionization of a previously electrically neutral molecule), estimated on the basis of the measurement of the photoelectron energy (PE), on the number of atoms in the alkane molecule.

This perhaps too lengthy and basically trivial example shows the advanced theoretical knowledge and research methods involved in explaining the structure and properties of hydrocarbons, a minor part of organic chemistry. Analogous and often much more advanced analyses concern more complex organic compounds, and based on them, we know a lot about the molecular structure and physicochemical properties, as well as the biological properties and effects, of macromolecules.

This also applies to macromolecules essential to living organisms, i.e. proteins, nucleic acids, polysaccharides, and lipids. Also in issues related to materials engineering, knowledge about the structural basics of materials is used in order to predict their potential properties. This is best seen in relation to carbon materials, where the same bonds between carbon atoms occur as in the hydrocarbon molecules discussed above.

sp^3 hybridization characterizes diamond, where there are only single bonds between carbon atoms, sp^2 is characteristic for graphene and graphite, where there are both single and double bonds, while linear carbon chains, called carbynes, are characterized by alternating triple and single bonds between carbon atoms, as in diacetylene. It is estimated that carbines are mechanically more durable than graphene and carbon nanotubes [28].

Mass spectrometry and photoelectron spectrometry are commonly used to identify these chemical compounds. So why is this type of theoretical analysis and identification by means of photoelectron spectra not applied to nanomaterials, trying to group them, for practical registration purposes, on the basis of methods that work well in the macroscopic world but not taking into account the internal structures and the presence of bonds?

5.5 More on biological material recognition

Since the components of biological macromolecules contain carbon atoms bonded to other atoms and atom groups, and are of nanometric size, they should also be considered as organic nanostructures (nanoparticles), thus the above considerations are also appropriate for biological macromolecules.

Exposing cells to nanostructures could result in a specific metabolic response to the stress caused by this contact [29]. This specificity results from the induction by a nanostructure, on the quantum mechanics level, of reversible changes in the spatial conformation of a protein capable of influencing the protein–protein interaction cascade and thus affecting cell signaling and the transcription process [30]. Thus the induction by a nanostructure of a new bonding orbital or the breaking of an existing bond by induction of an antibonding orbital in a receptor molecule can start a specific pathway of cellular metabolic response. Our experimental observations in this matter were the basis for patent [5] and was also discussed in our recent papers [6–8, 10, 31].

We have found that some genes respond in a similar manner after contact of a cell with any artificial material, whereas other genes respond in a specific manner to chemical composition and surface structure. Particularly important at this point is the question of whether the information capacity of the genome is sufficient for diversification of sources of stress. Assuming an analysis only within the 100 selected genes and a chosen 10 of them, the number of possible combinations is above 1×10^{13}, and for a choice of 50 out of 100 is much more than 10^{29}. The number of permutations of a set of 100 elements is equal to 100! which is more than 10^{158} combinations of the elements of this set. When one tries to analyze the full genome, the number of possible combinations is enormous.

The enormous information capacity of biological systems is exemplified by the specificity of antibody–antigen recognition process. Over the life of an organism the immune system learns to recognize and remember a huge number of antigens and specific antibodies are able to differentiate two very similar antigens. For example, monospecific polyclonal antibodies are able to recognize a specific antigen generated by the RGDS peptide present in a fibrinogen molecule, with almost no cross-reaction with the antigen presented by the RGDS peptide present in a fibronectin molecule [32].

In chapter 4, a microarray technique which may be suitable for the identification of products of nanotechnology is presented. Cells exposed to nanomaterials were lysed, and the RNA was isolated, amplified, and stained with a fluorophore. The labeled RNA was then applied to a whole genome microarray for hybridization with specific fragments of genes. Examples of transcriptome profiles, obtained for endothelial cells exposed to multi-walled carbon nanotubes (MWCNTs), silver nanoparticles (SNPs) and poly(amidoamine) dendrimers (PAMAM) were presented in figure 4.3. In another experimental work published by us recently [31], we analyzed the expression of genes of osteoblast cells (Saos-2) exposed to contact with samples made of the same Ti6Al4V titanium alloy, but prepared using different production methods. The samples were produced by CNC machining (ELI) or selective melting with electron beam technology (ELI-EBT). To facilitate the inference, an analysis of the relative changes in gene expression in cells exposed to the test samples in relation to the expression of these genes in cells grown under standard conditions (control) was performed. The fold change (FC) value was then taken as a measure of changes in gene expression. According to our criterion, significant changes in gene expression concerned those genes for which the $\log_2$ FC value was above 1 or below −1. Moreover, the criterion of the specificity of changes in gene expression was adopted, useful in differentiating the metabolic response of cells to contact with materials. Significant changes in FC values were observed only in response to contact of cells with one of the materials, while contact with the other type of material did not show significant changes in the expression of this gene or caused changes in the opposite direction, so these changes were defined as specific. The remaining significant changes, observed for both tested materials, were described as non-specific.

Figure 5.5 presents the profiles of changes in the expression of two hundred genes with numbers from 51 to 250 assigned to the microarray. For our purposes, the

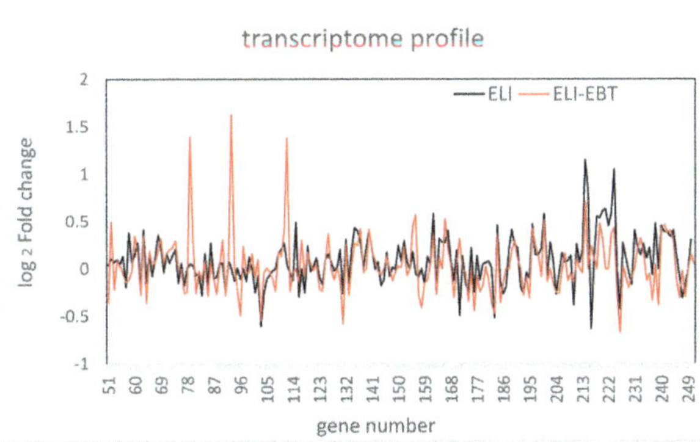

Figure 5.5. The gene expression profiles of 200 genes with the numbers assigned to the microarray in the range 51–251. These data allow us to compare the obtained gene expression profiles for cells exposed to samples made from the same titanium alloy Ti6Al7V but produced using different manufacturing methods.

names of these genes are not important, because we will not discuss the details of the metabolic pathways represented by these genes in terms of gene ontology analysis. We are instead interested in the similarities and differences in the expression patterns of these genes. It is easy to see that the similarity in expression concerns the vast majority of these genes, while differences occur only in a few cases. In accordance with the adopted criteria, we observed specific overexpression in cells in contact with the surface of the ELI-EBT sample for the genes numbered 79, 93, and 112, and non-specific overexpression for genes number 214 and 224 in cells in contact with the ELI sample.

Although individual genes are usually attributed to coding of appropriate peptides and proteins, a number of genes do not correlate with proteins, but contain a record of RNA molecules involved in the regulation of nuclear and cellular processes or perform other, so far undefined functions. Under the conditions of the fragment of our experiment described here, the *LOC400958, LINC01564*, and *CASC15* genes are located on the microarray at positions 79, 93, and 112, respectively.

The first two belong to the group called long intergenic non-protein coding RNA. The *LOC400958* gene shows functional associations with biological entities spanning several categories (molecular profile, disease, phenotype or trait, functional term, phrase or reference, cell line, cell type or tissue, gene, protein or microRNA), *LINC01564* is associated with activating the serine synthesis pathway and facilitating hepatocellular carcinogenesis, while the cancer susceptibility 15 (*CASC15*) gene produces a long non-coding RNA that can regulate cell proliferation.

The other two genes, *SNORA16A* (position 214) and *SNORD105B* (position 224) belong to the group known as small nucleolar RNAs and their function is to participate in the biogenesis (modification) of other small nuclear RNAs (snRNAs). This type of modifying RNA is located in the nucleolus of the eukaryotic cell which is a major site of snRNA biogenesis.

As can be seen, changes in the expression of these genes do not translate directly into changes in the proteome profile, but have an indirect effect.

Peptides and proteins, the products of the genes that are expressed in cells, also show a specific pattern in 2D electrophoresis, and also allow for identification of nanomaterials. Examples of 2D-DIGE separation of the proteome from endothelial cells exposed to the earlier above mentioned nanomaterials are presented in figure 4.4.

In the same work as above [31] we described changes in the proteome profile observed in cells grown on the surfaces of ELI and ELI-EBT samples. The analysis was performed using the 2D-DIGE electrophoresis technique [33].

As a result of the conducted separation in the gel, it was possible to find about 700 spots, of which for the sample marked as ELI there were upregulated 33 spots, while 190 spots were downregulated. For the sample marked as ELI-EBT it was 14 upregulated and 119 downregulated spots, respectively. In our hands only a small number of peptides and proteins with altered expression could be identified [31] by mass spectrometry [34]. Also in this group, there were non-specific changes (coding genes: *ACTB, RLA0, RSA*) and specific (coding genes: *LDHB, ENPL, ATPB/*

PDIA6, CH60/HNRPK, HNRPK, TBB5, TBA1B) defined in accordance with the above presented criterion.

The presented results strongly support the thesis that both the gene expression profile (transcriptome) and the profile of cellular proteins (proteome) have characteristic features that allow for the identification of the source of stress causing these changes, in this case contact with titanium alloy samples prepared with different technologies.

Since changes in the physicochemical properties of nanomaterials, in relation to the properties of these materials in macroscopic form, do not change by leaps and bounds, but rather exhibit the nature of the functional dependency [35], it can be assumed that these functional relationships for different nanomaterials may have a common part (curves intersect or overlap each other). The changes mentioned above translate into changes in the electronic structure of nanomaterials, which allows the introduction of nanodescriptors in the nano-QSAR analysis to predict the properties of nanoparticles [36].

Taking into account the above, one should bear in mind that the same, or very similar, physicochemical properties may be present in nanomaterials of different chemical composition and different surface structure, if the structure of the electronic configuration on the surface of these objects is the same or very similar. This is an inference analogous to the case of single atoms, where the electronic structure of the atom determines its physicochemical properties, more precisely the electronic structure of the last valence electron levels, and, going further, the electronic structure of electrons available on the surface of molecules determines most of the physicochemical properties of the molecules. Performing the next step, we should introduce the concept of the orbital–electron energy pair instead of the electronic structure. This pair clearly defines the properties of the atom, molecule or nanostructure in relation to that of electrons available on the surface. Considering the fact that the sizes of the molecules involved in the biochemical activity of cells are also within the nanometric scale, it is difficult not to also take into account the above considerations for them. However, it should be taken into account that these are nanoparticles with a complicated hetero-atomic chemical composition and a complicated spatial structure implicating the complex electronic structure of the surface. This results in a variety of surface orbitals and corresponding energies available on the surface, which implies the participation of these bio-nanoparticles in many different but highly specific interactions. Experimental data seem to confirm this course of thought. The interaction of engineered nanoparticles with cells is strongly dependent on their size and shape [37–40], and thus also on the electronic surface structure of these nanomaterials, in other words it depends on the shape of the orbitals and the corresponding electron energies. It is a decisive factor conditioning the specificity of the interaction of nanomaterials with cells and is responsible for the highly specific metabolic response of a cell exposed to contact with these nanomaterials.

5.6 Some practical considerations

It is worth considering the practical aspect of the proposed solutions. Despite the previously postulated possibility of imaging electron orbitals available on the nanoparticle/nanostructure surface, the limitations of the currently available methods and their labor consumption should be taken into account.

Numerous and long conversations with practitioners in the field of scanning tunneling microscopy (STM) reveal some important difficulties that need to be overcome. First, to develop a procedure to obtain the conditions required to achieve the proper spatial orientation of the nanoparticle and to reproduce such an orientation. Second, the development of standards for dealing with nanoparticles of various spatial structures. After overcoming these difficulties, there is another unknown about the criteria for comparing the similarities and differences of orbital images, which images for complex nanoparticle structures will also be complicated, more complicated than the images of atomic orbitals and simple molecular orbitals.

Without prejudging the possibility of carrying out such analyses, it seems that it will be much easier to stop, at least for the moment, on the analysis of the energy of the eigenstates (self-energies) of the available electrons, i.e. the second component of the pair orbital–self-energy of the eigenstate.

In this case, using photoelectron spectrometry (PES) methods it is possible to obtain a global photoelectron energy spectrum from all excited atoms in the nanoparticle. Of course, the complexity of this spectrum will depend on the nanoparticle size, atomic composition, type of bonds between atoms, stabilizer or nanoparticle surface modification, energy of radiation used, homogeneity and spatial orientation of nanoparticles, and many other factors not mentioned here. Such an image is highly disturbing for a researcher embedded in basic research and focused on establishing the properties of the tested sample on the basis of the obtained spectrum. However, for our purposes, the level of spectral complication is not so worrying, because we want to remain at the level of comparing the similarities and differences of spectra obtained for different samples of nanomaterials and, on this basis, obtain the possibility of distinguishing nanomaterials in order to identify them. By collecting such an averaged global spectrum from a large population of nanoparticles, we will become independent of the spatial orientation and, to a large extent, of the homogeneity of nanoparticles. However, using X (XPS) or UV (UPS) radiation we will obtain information coming from the inside (nanoparticle core) using XPS, or from its surface using UPS.

With regard to molecular biology methods, it should be noted that currently available methods significantly accelerate the research process. Instead of transcriptome analysis using the microarray technique, it would be more productive to use next-generation sequencing systems (NGS) [41], and instead of proteome analysis using gel electrophoresis, gel-free methods based on a combination of liquid chromatography with mass spectroscopy (LC-MS/MS) known as the shotgun technique are more efficient [42].

The analysis of the similarities and differences of photoelectron spectra as well as transcriptome and proteome profiles also requires some comment. Various strategies for this analysis are possible. Presented below are our own preliminary results of the analysis of the similarities and differences of fragments of the transcriptome profiles of EA.hy926 cells exposed to contact with surfaces of various biomaterials (figure 5.6).

Correlation coefficient

	AISI316L	*PEEK*	*ELI*	*ELI-EBT*	*VK100*	*MENDEC*
AISI316L	1					
PEEK	0.840083	1				
ELI	-0.01266	-0.03633	1			
ELI-EBT	-0.03253	-0.05938	0.952547	1		
VK100	-0.10622	-0.10934	0.024779	0.025347	1	
MENDEC	0.328975	0.281401	-0.12452	-0.12918	-0.14577	1

AISI316L - medical steel
PEEK - medical grade polymer
ELI - medical grade titanium alloy Ti6Al4V
ELI EBT - as above, but produced by 3D printing
VK100 - medical grade cement for surgery
MENDEC - medical grade cement for surgery

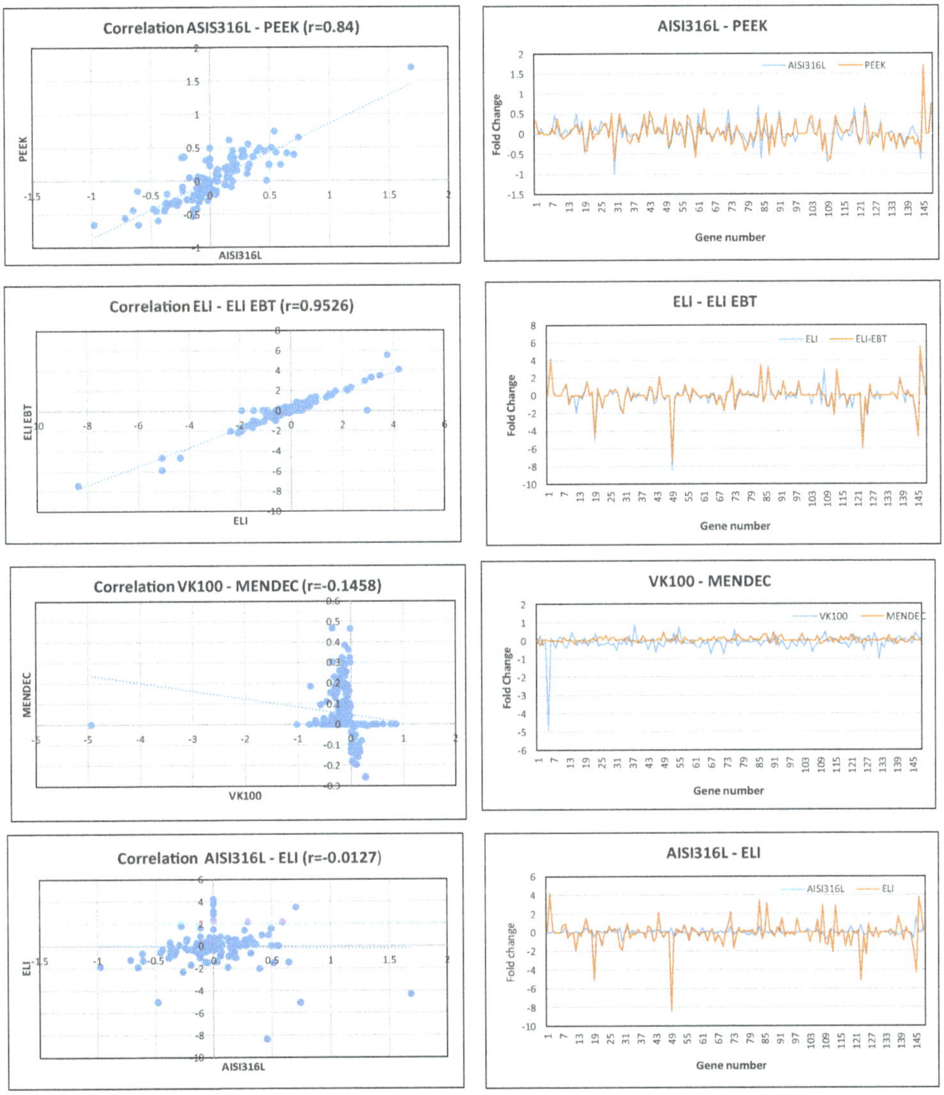

Figure 5.6. The table at the top of the figure contains results of the correlation analysis of gene expression profile fragments (147 genes in the order as on the microarray) of EA.hy926 cells grown in contact with the surfaces of medical steel AISI316L, medical grade polymer PEEK, medical grade titanium alloy Ti6Al4V machined (ELI), and the same titanium alloy Ti6Al4V 3D printed (ELI-EBT), bone surgery cements VK100 (silicon polymer), and MENDEC (polymethyl methacrylate). Graphical representations of the correlation (left side) and gene expression profiles (right side) are also presented.

In the specific case of two samples with the same chemical composition, but with a different structure of the surface topography (cutting versus electron beam sintering in a 3D printer), a very high level of correlation $r > 0.9527$ was obtained with very similar gene expression profiles. This means that EA.hy926 cells are capable of recognizing the two samples as very similar, but not identical. In repeated tests for pairs of identically prepared samples, the correlation coefficient exceeded the value of 0.99, and the gene expression profiles were practically indistinguishable. In the case of other biomaterials, the correlation of gene expression profiles was much lower, which was reflected in the graphically presented overlapping of transcriptome profiles (figure 5.6). These preliminary observations strongly suggest the ability of the cell to specifically recognize the stress resulting from contact with materials that differ in both chemical composition and surface structure, and thus indicate the possibility of using these skills of the cell and apply it as a specific sensor recognizing the material and its surface structure.

We believe that collecting experimental data of transcriptome and proteome profiles in a database will allow for specific identification of materials or nanomaterials, similarly to the identification of chemical substances based on the specific spectrum of mass spectrometry and comparing it with the data available in existing databases. The first very small step has already been made by creation of the cBiT database—a transcriptomics database for innovative biomaterial engineering [13].

Concluding the above, the concept of an innovative strategy of nanomaterial identification has been developed to meet the needs for efficient classification and grouping of nanomaterials for proper risk assessment and management with the use of nano-QSAR modeling and read-across approach. It relies on a combination of already available methods, involving the significance of quantum effects of nanomaterials and specific cell's metabolic response upon contact with nanomaterials, and indicates a theoretical and experimental approach making this identification possible. We believe that the near future will show that the herein proposed new concept is operational and useful.

References

[1] Oomen A G *et al* 2015 Grouping and read-across approaches for risk assessment of nanomaterials *Int. J. Environ. Res. Public Health* **12** 13415–34

[2] Bos P M J *et al* 2015 The MARINA risk assessment strategy: a flexible strategy for efficient information collection and risk assessment of nanomaterials *Int. J. Environ. Res.* **12** 15007–21

[3] Trump B D, Hristozov D, Malloyd T and Linkov I 2018 Risk associated with engineered nanomaterials: different tools for different ways to govern *Nano Today* **21** 9–13

[4] Ghosh C K 2015 Quantum effect on properties of nanomaterials *Introduction to Nano* ed A Sengupta and C K Sarkar (Berlin: Springer) pp 73–111

[5] Walkowiak B, Komorowski P and Walkowiak-Przybylo M 2012 Means of identification of materials engineering products or nanotechnologies *PL Patent* PAT.223224

[6] Komorowski P *et al* 2019 Simultaneous transcriptome and proteome analysis of EA.hy926 cells under stress conditions induced by nanomaterials *J. Biomed. Mater. Res.* B **107** 1024–34

[7] Komorowski P, Walkowiak-Przybylo M and Walkowiak B 2016 Early cell response to contact with biomaterial's surface *J. Biomed. Mater. Res.* B **104** 880–93

[8] Walkowiak-Przybylo M, Komorowski P and Walkowiak B 2017 Differences in the expression of cell cycle genes in osteoblasts and endothelial cells cultured on the surfaces of Ti6Al4V and Ti6Al7Nb alloys *J. Biomed. Mater. Res.* A **105** 1607–17

[9] Ho Kim J *et al* 2016 *C. elegans*-on-a-chip for *in situ* and *in vivo* Ag nanoparticles' uptake and toxicity assay *Sci. Rep.* **7** 40225

[10] Walczynska M, Jakubowski W, Wasiak T, Kadziola K, Bartoszek N, Kotarba S, Siatkowska M, Komorowski P and Walkowiak B 2018 Toxicity of silver nanoparticles, multiwalled carbon nanotubes, and dendrimers assessed with multicellular organism *Caenorhabditis elegans Toxicol. Mechan. Methods* **28** 432–9

[11] Cranford S and Buehler M J 2010 Materiomics: biological protein materials, from nano to macro *Nanotechnol. Sci. Appl.* **3** 127–48

[12] Groen N, Yuan H, Hebels D G A J, Koçer G, Mbuyi F, LaPointe V, Truckenmüller R, van Blitterswijk C A, Habibović P and de Boer J 2017 Linking the transcriptional landscape of bone induction to biomaterial design parameters *Adv. Mater.* **29** 1603259

[13] Hebels D G A J, Carlier A, Coonen M L J, Theunissen D H and de Boer J 2017 cBiT: A transcriptomics database for innovative biomateriał engineering *Biomaterials* **149** 88–97

[14] Toropova A P, Toropov A A, Veselinović A M, Veselinović J B, Benfenati E, Leszczynska D and Leszczynski J 2016 Nano-QSAR: model of mutagenicity of fullerene as a mathematical function of different conditions *Ecotoxicol. Environ. Saf.* **124** 32–6

[15] Zhu H *et al* 2016 Supporting read-across using biological data *ALTEX* **33** 167–82

[16] Gold *Causes of Color* http://www.webexhibits.org/causesofcolor/9.html

[17] Scholl J A, Koh A L and Dionne J A 2012 Quantum plasmon resonances of individual metallic nanoparticles *Nature* **483** 421–7

[18] Agrawal Y, Kedawat G, Kumar P, Dwivedi J, Singh V N, Gupta R K and Gupta B K 2015 High-performance stable field emission with ultralow turn on voltage from rGO conformal coated TiO_2 nanotubes 3D arrays *Sci. Rep.* **5** 11612

[19] Ramos M *et al* 2013 Hardness and elastic modulus on six-fold symmetry gold nanoparticles *Materials* **6** 198–205

[20] Gao F and Gu Z 2016 Melting temperature of metallic nanoparticles *Handbook of Nanoparticles* ed M Aliofkhazraei (Cham: Springer)

[21] The orbitron *Chemistry Nexus* https://www.webelements.com/nexus/tag/orbitron/

[22] *Orbital Viewer* https://orbitals.com/orb/ov.htm

[23] Cheng Z, Du S, Guo W, Gao L, Deng Z, Jiang N, Guo H, Tang H and Gao H-J 2011 Direct imaging of molecular orbitals of metal phthalocyanines on metal surfaces with an O_2-functionalized tip of a scanning tunneling microscope *Nano Res.* **4** 523–30

[24] Offenbacher H, Lüftner D, Ules T, Reinisch E M, Koller G, Puschnig P and Ramsey G 2015 Orbital tomography: molecular band maps, momentum maps and the imaging of real space orbitals of adsorbed molecules *J. Electron Spectrosc. Relat. Phenom.* **204** 92–101

[25] McMurry J 2011 *Fundamentals of Organic Chemistry* 7th edn (Belmont, CA: Brooks/Cole)

[26] NIST 2022 NIST Chemistry WebBook *NIST Standard Reference Database* Number 69 National Institute of Standards and Technology

[27] Amador-Muñoz O, Misztal P K, Weber R, Worton D R, Zhang H, Drozd G and Goldstein A H 2016 Sensitive detection of n-alkanes using a mixed ionization mode proton-transfer-reaction mass spectrometer *Atmos. Meas. Tech.* **9** 5315–29

[28] Liu M, Artyukhov V I, Lee H, Xu F and Yakobson B I 2013 Carbyne from first principles: chain of C atoms, a nanorod or a nanorope *ACS Nano* **7** 10075–82

[29] Samant M P, Lahir Y K, Chitre A V and Kale A 2018 Review: influence of nanomaterials on cellular stress, and cellular behavior *Adv. Clin. Toxicol.* **3** 000140

[30] Mukhopadhyay A, Basu S, Singha S and Patra H 2018 Inner-view of nanomaterial incited protein conformational changes: insights into designable interaction *AAAS Res.* **2018** 9712832

[31] Komorowski P *et al* 2020 Comprehensive biological evaluation of biomaterials used in spinal and orthopedic surgery *Materials* **13** 4769

[32] Walkowiak B, Michalak E, Borkowska E, Koziolkiewicz W and Cierniewski C S 1994 Concentration of RGDS-containing degradation products in uremic plasma is correlated with progression in renal failure *Thromb. Res.* **76** 133–44

[33] GE Healthcare 2005 *Ettan DIGE System User Manual* GE Healthcare https://www.mcgill.ca/cian/files/cian/ge_dige_manual.pdf

[34] Bantscheff M, Schirle M, Sweetman G, Rick J and Kuster B 2007 Quantitative mass spectrometry in proteomics: a critical review *Anal. Bioanal. Chem.* **389** 1017–31

[35] Guisbiers G 2010 Size-dependent materials properties toward a universal equation *Nanoscale Res. Lett.* **5** 1132–6

[36] Jagiello K *et al* 2017 Size-dependent electronic properties of nanomaterials: how this novel class of nanodescriptors supposed to be calculated? *Struct. Chem.* **28** 635–43

[37] Neshatian M, Chung S, Yohan D, Yang C and Chithrani D B 2014 Determining the size dependence of colloidal gold nanoparticle uptake in a tumor-like interface (hypoxic) *Colloids Interface Sci. Commun.* **1** 57–61

[38] Shang L, Nienhaus K and Nienhaus G U 2014 Engineered nanoparticles interacting with cells: size matters *J. Nanobiotechnol.* **12** 5

[39] Khan I, Saeed K and Khan I 2017 Nanoparticles: properties, applications and toxicities *Arab. J. Chem.* **12** 908–31

[40] Akter M, Sikder M D T, Rahman M D M, Atique Ullah A K M, Hossain K F B, Banik S, Hosokawa T, Saito T and Kurasaki M 2018 A systematic review on silver nanoparticles-induced cytotoxicity: physicochemical properties and perspectives *J. Adv. Res.* **9** 1–16

[41] Slatko B E, Gardner A F and Ausubel F M 2018 Overview of next generation sequencing technologies *Curr. Protoc. Mol. Biol.* **122** e59

[42] Tyanova S, Temu T and Cox J 2016 The MaxQuant computational platform for mass spectrometry-based shotgun *Nat. Protoc.* **11** 2301–19

IOP Publishing

Biological Evaluation of Materials
The interaction of materials with their environment
Bogdan Walkowiak, Magdalena Walkowiak-Przybyło and Piotr Komorowski

Chapter 6

Considerations of substrate bonding and catalysis models

There seems to be no need to convince anyone of the importance of chemical reactions. This chapter presents the basic information on the kinetics of reactions and the kinetic theory of collisions, as well as the basics of catalysis. The possibility of lowering the activation energy of a reaction based on the possibility of tunneling the potential barrier by electrons is explained. The possibility of tunneling of barriers by protons is also indicated, which enables the functioning of hydrogen bonds and the occurrence of numerous enzymatic reactions. The possibility of creating tunnel structures enabling control over the transfer of multiatomic structures in enzymatic chains is also indicated. The essence of the influence of weaker energy occurring in a huge number of electrostatic interactions on molecular stability and the processes of molecular recognition are indicated. These properties are taken into account in the case of interactions of nanoparticles with protein particles.

6.1 The essence of the chemical reaction

The mystery of combining atoms into molecules, which is the essence of chemical reactions, despite the existence of numerous studies explaining this process, still remains an exciting and not always clear act of creation of a new substance. This is particularly important in relation to living organisms, the existence and functioning of which are based entirely on dynamically running processes of molecular interactions leading to the formation, transformation, or decay of the macromolecules essential for life. This interaction is carried out by creating new bonds between atoms and groups of atoms in order to create a new or modified molecule of different quality from the original molecule, or by breaking existing bonds in order to degrade complex molecules into simple products in order to eliminate them or use them to create new molecular structures. In the case of inanimate matter, the formation and breaking of bonds between atoms, although also extremely important

doi:10.1088/978-0-7503-2656-8ch6

in the process of creating new substances and their durability, has a different meaning due to the relatively long durability of the resulting materials or their degradation products.

There is no doubt that the necessary condition for any interaction of atoms with molecules and molecules to occur with each other is their existence and the possibility of meeting at a distance enabling this interaction. This condition is just as critical in classical physics and chemistry, where, in simplified terms, one can adopt a deterministic approach for the interpretation of processes, as well as in physics and quantum chemistry, where indeterminism rules by definition. In the latter case, the wave function for a given particle must be normalized by superimposing the certainty of the particle's existence. However, the condition of meeting reactants (substrates) at a distance enabling interaction, intuitively obvious in the classic description of the kinetics of a chemical reaction, does not necessarily have to be required in a quantum mechanical description due to the recently demonstrated existence of quantum entanglement [1, 2]. Let us begin our considerations with the classical approach to the kinetics of chemical reactions and its development to enzymatic reactions.

6.2 Reaction kinetics

Each act of reaction is a complex process of creating new bonds preceded by breaking existing bonds. Therefore, the speed of a particular reaction depends on the nature of the existing and emerging bonds participating in that reaction. Regardless of this, a number of environmental factors have a significant influence on the speed of the reaction. In the first place, these are the availability of the reactants, expressed by their concentration in the reaction medium, and the kinetic energy of the reactants, expressed by the temperature of the reaction medium.

For the sake of narrative order, it is worth recalling the order of the reaction, which is defined as the sum of the exponents of the powers in the kinetic equation presented in the form of a power monomial:

$$r = k[\mathrm{A}]^n[\mathrm{B}]^m[\mathrm{C}]^p[\mathrm{D}]^q\ldots \tag{6.1}$$

where r is the reaction rate, k is the reaction constant, [A], [B], [C], and [D] are the molar concentrations of the reactants, while the exponents n, m, p, and q are real numbers indicating the power to which the concentration of a given reagent should be increased to satisfy the conditions of the kinetic equation.

The value of n determines the order of the reaction with respect to reagent A, m with respect to reagent B, and so on. In this sense, the following example provisions should be given specific meanings:

$$r = k[\mathrm{A}]^0 = k - \text{(zero-order reaction)} \tag{6.2}$$

$$r = k[\mathrm{A}]^1 = k[\mathrm{A}] - \text{first order reaction} \tag{6.3}$$

$$r = k[\mathrm{A}]^2 - \text{second order reaction} \tag{6.4}$$

$$r = k[\mathrm{A}][\mathrm{B}] \text{ – second-order reaction, but first-order reaction in respect to components A and B} \tag{6.5}$$

$$r = k[\mathrm{A}]^{1,5} \text{ – fractional order reaction (complex reaction mechanism).} \tag{6.6}$$

The rate of the zero-order reaction does not depend on the concentration of the reagent—for example, the photochemical reaction resulting from the absorption of a photon is not dependent on the concentration of the absorbing molecules. On the other hand, the rate of the first-order reaction depends only on the concentration of one reagent and decreases exponentially with time with the decrease of the concentration of this reagent. An example of a first-order reaction would be the disappearance of the fluorescence of an excited molecule with fluorescent properties. The first-order reaction does not depend on the probability of collisions of the reagent molecules, but may depend on the temperature of the reaction environment. An example of a first-order pseudo-reaction would be the reaction against one of the reagents in the case of high excess in the concentration of the second reagent, then the rate of the reaction also depends only on the initial concentration of the first reagent and the duration of the reaction.

The second-order reaction is considered to be the primary chemical reaction in which the rate of the reaction depends on the probability of the meeting of the reagents, i.e. the concentration of these reagents. In the case of homogeneity of the reagents, for example during the formation of H_2 or O_2 molecules, the rate of the reaction over time depends on the initial concentration of the individual hydrogen or oxygen atoms and the duration of the reaction, i.e. the depletion of the reagents. In the case of heterogeneous reactions, when we deal with different reagents denoted as A and B, the rate of reaction in time depends on the product of the concentrations of both reagents and also changes with the depletion of the reagents. Higher order reactions requiring more than two reagents to meet at one point in space at the same time are relatively rarely observed due to the difficulty of meeting the condition of unity of place and time by more than two reagents. Usually, we observe follow-up reactions, i.e. a second-order reaction produces a product that is also a reactant in the next step reaction.

It has been experimentally found that an increase in temperature contributes to an increase in the rate of the reaction, be it endo- or exothermic. This change is contained in the reaction rate constant:

$$k = A\exp(-E_{\mathrm{A}}RT) \tag{6.7}$$

where A is the characteristic constant for a given reaction, E_{A} is the energy of reaction activation, T is the temperature on the absolute scale (K), and R is the universal gas constant equal to 8.31 J mol^{-1} K^{-1}.

The dependence of the reaction rate constant k on the temperature T is exponential, which explains the significant changes in the reaction rate with relatively small changes in temperature.

6.3 Kinetic theory of collisions

Many phenomena concerning the kinetics of reactions can be successfully explained on the basis of the kinetic theory of collisions. According to the basic assumptions of this theory, a reaction can take place when the molecules of the reactants collide and as a result of this collision the atoms are rearranged, the existing bonds are broken and new bonds are formed, resulting in the formation of a reaction product. The collision theory assumes that the speed of each reaction step is proportional to the number of collisions of molecules involved in a given step per unit time and depends on the ratio of the number of effective collisions to the number of all collisions.

Obviously, not all collisions are effective and result in a reaction. The effectiveness of the meeting (collision) is determined by both the kinetic energy of the moving regents and the repulsive effect of the negative electric charges of the electrons of the colliding reactants, the minimum energy necessary for the reaction to take place is called the activation energy E_A. Reactions requiring a high activation energy usually take place slowly, because at a given temperature only some of the reagent molecules have sufficient energy during the collision to overcome the energy barrier determined by the activation energy. On the other hand, reactions characterized by low activation energy run faster, because a greater percentage of the reactant molecules have sufficient kinetic energy for overcoming activation energy and take place of the reaction. It should also be noted that as a result of some reactions, called spontaneous reactions, excess free energy is released in the form of heat (exothermic reaction), while for other reactions it is necessary to supply free energy (endothermic reaction). Such situations are illustrated in figure 6.1.

According to the kinetic collision theory, only those reactants whose total energy is higher than the activation energy E_A can carry out a chemical reaction, and the activation energy itself is a kind of energy barrier that limits the speed of the reaction.

6.4 Catalysis and tunneling

The speed of the reaction can be increased by the use of a catalyst (in biology, a catalyst is called an enzyme), which is capable of increasing the number of effective collisions as a result of lowering the activation energy of the reaction. It should be noted that the catalyst is not consumed in the catalyzed reaction (there is no loss and no more catalyst) and therefore is not present in the kinetic equation either on the reagent side (substrate) or on the product side. Figure 6.2 shows changes in the activation energy value of a given reaction due to the presence of a catalyst. The uncatalyzed reaction requires overcoming the activation energy of E_{A1}, while the presence of the catalyst lowers this energy to the value of E_{A2}, thereby increasing the rate of the reaction. When the catalyst is able to form intermediates with reactants, we deal with catalysis of the reaction through intermediates, the formation of which requires even lower activation energies of E_{A3} and E_{A4}, which leads to an even more effective increase in the overall reaction speed.

In light of the success of the quantum mechanical description of the atom and the application of this description in relation to the formation of molecular bonds, it is

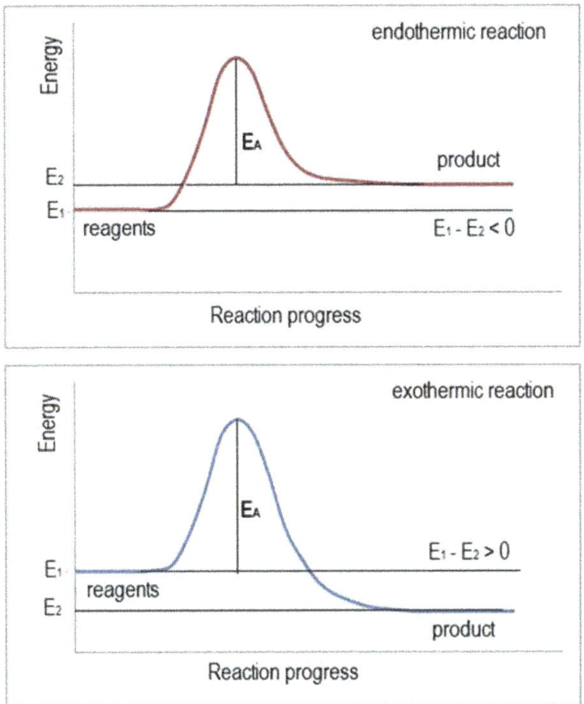

Figure 6.1. Schematic representation of the activation energy (E_A) as an energy barrier to the reactions taking place. The difference in the energy of the products in relation to the energy of the reagents in the endothermic and exothermic reactions is also shown schematically.

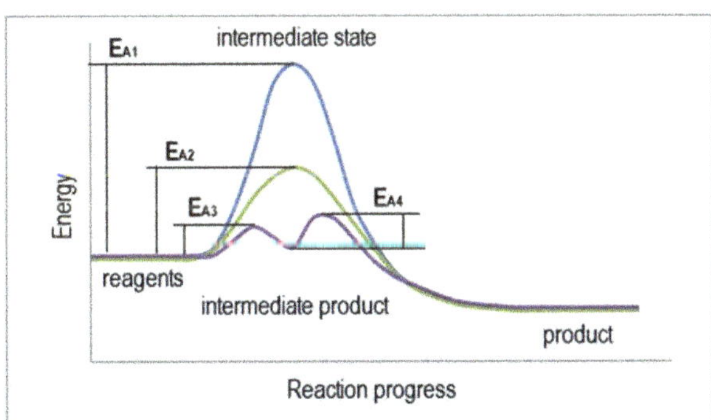

Figure 6.2. Schematic representation of the activation energy (E_A) as an energy barrier to the reactions taking place. E_{A1}—activation energy of the uncatalyzed reaction; E_{A2}—catalyzed reaction activation energy; E_{A3} and E_{A4}—activation energies of the reaction in which intermediate products participate.

worth paying attention for a moment to the process of tunneling the energy barrier (potential barrier) limiting the effectiveness of the collision, which is quite common in the atomic, subatomic, and molecular scales, but not observed in the macroscopic scale. This absence of the macroscopic tunneling process (we do not observe the penetration of objects through walls) does not, however, limit the effects of this process taking place on the atomic and molecular scale, and observed on the macroscopic scale, for example α-decay occurring in the atomic nucleus is observed on the macroscopic scale as α radiation.

Let us look at the nuclear reaction in α-decay. The α particle, which is the nucleus of the helium atom (two protons and two neutrons), is emitted from a large unstable atomic nucleus, carrying a strictly defined kinetic energy. There would be nothing strange in this, if not for the fact that the energy of α particles emitted as a result of the decay of the nucleus is much lower than the binding energy of these nucleons in the nucleus, represented by the energy barrier of nuclear bonds. Classical consideration of this process requires that the system of these four nucleons, forming as a result of decay of the helium nucleus, obtain energy greater than or equal to the activation energy (potential barrier energy) in order to be able to leave the decaying nucleus, so after decay the α particle should have this excess of energy in form of kinetic energy. It would have to be concluded that such a reaction cannot occur or takes place at a very slow rate. However, the experimental data do not confirm such a high kinetic energy of α particles after decay and indicate significant reaction rates. Quantum physics provides a relatively simple, but still hard to accept for many, solution in the form of the process of tunneling of the potential barrier. Figure 6.3 shows the idea of the tunneling process schematically.

In quantum mechanics, according to the wave–particle duality theorem, each particle can exhibit corpuscular or wave characteristics, depending on the conditions of the experiment. Let us assume that we assign wave properties to the particle α considered here and describe them with the wave function $\Psi(x)$ representing the probability of finding the particle in a given region of space. In figure 6.3 the interior space of the nucleus is represented on the left of the potential barrier formed by the nuclear bond forces. In the area of the nucleus, we observe a high amplitude of the wave function, which corresponds to a high probability of finding the nucleotides that make up the α particle just inside the nucleus. However, the probability of finding the α particle inside the potential barrier is not zero and depends on the

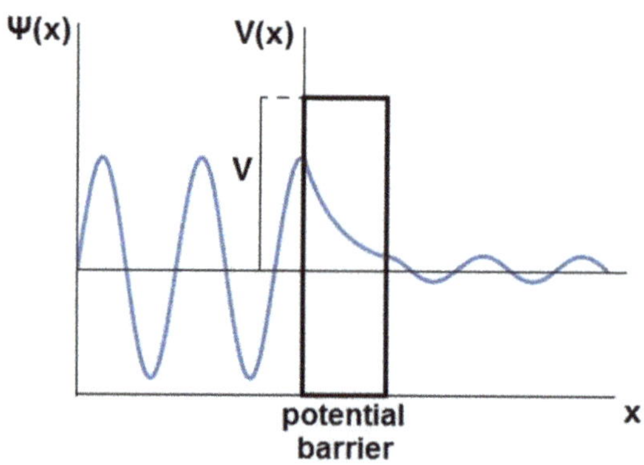

Figure 6.3. Schematic representation of the idea of tunneling of a potential barrier of height V and finite thickness by a particle described by the wave function $\Psi(x)$.

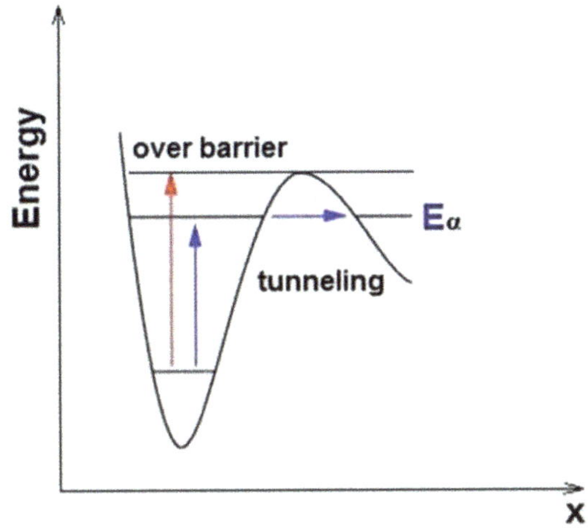

Figure 6.4. Schematic representation of the potential barrier tunneling process by the α particle in the nuclear decay process.

distance from the nucleus boundary and the width of the barrier. This dependence is illustrated by the decreasing wave function curve inside the barrier. Since this barrier has a finite width, the probability of finding a particle α outside this barrier on the right-hand side is also not zero, although much lower than inside the nucleus and in the region of the potential barrier, which is shown as a significantly lower amplitude of the wave function. At the boundaries of the potential barrier, the wave function on the left- and right-hand side of each boundary must assume the same values, ensuring the continuity of the wave function. This non-zero probability of finding the α particle in the potential barrier region represents the tunneling process of this barrier by the considered α particle, and, consequently, the penetration of α particles with energies significantly lower than the activation energy (nucleon binding energy in the nucleus) through this barrier, which is shown schematically in figure 6.4.

The process of tunneling by the α particle of the barrier created by nuclear forces is not isolated, but rather it should be stated that the tunneling process of energy barriers created by various types of forces binding atoms and groups of atoms in molecules is common and is responsible for the transfer of electrons and protons during the formation and breaking of chemical bonds in molecules. The α decay reaction was mentioned here due to its high visibility, the universality of the described phenomena, and their educational potential [3], while the tunnelling effect itself is certainly invaluable not only in physics and chemistry, but also in biology and seems to be an important element of evolution the world [4].

6.5 Chemical bonds

Chemical bonds, although of a completely different nature from nuclear bonds, are also formed by overcoming the activation energy of the reaction, which may occur when the reactants have a sufficiently high energy and cross the barrier, but can also be formed at lower reactant energies by tunneling of the barrier represented by activation energy.

Bonds, classified into two categories, are important for the formation and stability of a chemical compound. These categories are covalent (atomic) and ionic bonds. The remaining bonds, resulting from the electrostatic interaction of the charges present in the molecule or induced in it, are much weaker, but occurring en-mass they are of great importance in maintaining the spatial structure of the molecule and creating molecular interactions. Let us begin our considerations with covalent bonds, considered to be the strongest chemical bonds present in molecules.

Free, single atoms are extremely rare. The noble gas atoms most often present in atomic form do not follow this trend, which is traditionally explained by the filling of the valence shell by two (helium—electron doublet) or eight electrons (neon, argon, krypton, xenon, radon—electron octet). The remaining atoms are usually bound in smaller or larger molecules of chemical compounds. This bond is most often carried out by the sharing of an electron pair by atoms that make up a given chemical compound. Considering the H_2 molecule formed by the two simplest hydrogen atoms, each of which has one electron, we can indicate the commonality of these electrons and then each of the two atoms has an electron doublet. Thanks to this electron sharing, each of the hydrogen atoms obtains the electronic configuration of the helium atom, and the resulting molecule is stable. It is worth considering how such a bond is implemented. According to the classical theory of electrostatic interactions, the positively charged nucleus of the hydrogen atom H^+ strongly attracts the negatively charged electron, and the strength of this interaction increases as the distance between the charges decreases, which should cause the electron to drop to the nucleus. The classical theory of the Bohr atom eliminated this effect by introducing stable electron orbits for which the quantization condition of the angular momentum of the electron in the atom is met. Seemingly, this resolved the problem of the stability of the atom, but did not answer why, or rather how, a bond arises between two identical atoms. From the point of view of quantum mechanics, in an electrically neutral hydrogen atom the electron also tends to be as close as possible to the nucleus having a positive electric charge, but the Heisenberg uncertainty principle imposes a condition requiring an increase in the total energy of the electron with a decrease in the volume occupied by this electron, which means that in direct contact of the electron with the nucleus, the total energy of the electron would have to be infinite. This places a limit on the volume of the 1s orbital for an electron in a hydrogen atom. Because, according to the assumptions of quantum mechanics, we can assign an electron a wave function that determines the probability of finding this electron in the region of space we are interested in, the probability of finding an electron belonging to one of the hydrogen atoms in a nearby other hydrogen atom is not zero, and this means the possibility of sharing electrons and the formation of a covalent bond between two hydrogen atoms and the formation of an H_2 molecule. In the resulting molecule, the volume of the binding orbital resulting from the superposition and hybridization of the orbitals of two electrons from two hydrogen atoms is greater than the volume of the initial orbitals, which means lower total energy of both electrons in the H_2 molecule compared to the energies in separate free atoms. However, attention should be paid to the fact that when creating such a bond, all the rules of quantum mechanics must be met, including the Pauli principle exclusion in one quantum object to occupy two energy levels described with the same quantum numbers. In the case of the formation of a covalent bond between the hydrogen atoms, the electrons forming this bond differ in spin. For this reason, it is not possible to create such a molecule from two helium atoms, because in each of the helium atoms there are already two electrons with opposite spins [5, 6].

Similarly, the formation of a covalent bond in all other diatomic molecules made of the same atoms (homonuclear molecules) can be explained in a similar way. Although during the formation of a covalent bond between two different atoms, the hybridization of orbitals and the formation of binding orbitals are similar, using the tunneling effect, the resulting bond is slightly different from that in the homonuclear system. To put it in simply language, in homonuclear molecules the shared electrons reside in each of the atoms with the same probability, while in diatomic molecules of different atoms (hetero-nuclear) the probability is different. The atom that binds the shared electron more strongly (called the electronegative atom) presents the conditions for the electron to stay in it with a greater probability than in the electropositive atom. The resulting bond is not symmetrical and becomes polarized. The greater the probability of finding shared electrons in one of the atoms (more electronegative), the more polarized the bond is, until the electronegative atom takes both electrons and becomes a negative ion (an excess of one electron), and the other atom becomes a positive ion (deficiency of one of the electrons). In this way, an ionic bond is formed, equally common in the formation of molecules, although less durable due to the possibility of decay under the influence of polar environmental molecules (dissociation). It is extremely difficult to indicate the point at which a polarized covalent bond becomes an ionic bond. Currently, the classification of a bond to the group of covalent bonds is determined by the results of electron density measurements in the area between atoms, while according to the criteria of quantum chemistry, it can be assumed that an ionic bond takes place when the theoretically calculated energy of the bonding molecular orbitals is greater than the ionization potential of the electron donor atom. In such a situation, the formation of an ionic bond is energetically more advantageous.

Theoretical considerations show that even a slight overlap of electron orbitals may be sufficient to transfer an electron and create a bond [7]. Until recently, it was commonly believed, in line with the kinetic collision theory, that the reactants must collide in order for a reaction to take place. Meanwhile, electrons are able to travel long distances, even counted in nanometers, in the tunneling process, thus enabling the reaction to take place without the necessity of a collision of the reactants, but the distance influences the rate of reaction [8, 9].

The tunneling of the potential barrier by electrons enables a variety of biological processes to take place, for example electrons can be moved (transported) through connections between proteins [10].

It follows that the phenomenon of tunneling, common in the subatomic, atomic, and molecular scales, enables the formation of bonds without the need for collisions of reagents and seems to be universally used in the course of chemical reactions [11], although this conclusion should not be generalized, as some observations suggest that the kinetics of electron transfer over long distances between the protein partners of the redox reaction in the aqueous environment differs from the expected electron tunneling kinetics [12].

6.6 Hydrogen bonds and electrostatic interactions

Another common bond, responsible mainly for the stabilization of the spatial structure of molecules and their direct interaction, is hydrogen bonding. This bond is schematically represented as (X)Dn–H···Ac(Y–Z), where H is a hydrogen atom, (X)Dn is a hydrogen bond donor represented by an X atom, and Ac is its acceptor represented by a Y atom linked covalently with the Z atom. The solid line represents a polarized covalent bond, while the dashed line represents a hydrogen bond. The diagram showing this binding indicates a high degree of complication of the mechanism of this binding. It is necessary that the hydrogen atom is covalently bonded to another atom or a group of atoms with high electronegativity, thus gaining an excess of positive electric charge. On the other hand, the acceptor atom must also be electronegative to present a negative electric charge enabling electrostatic interaction with an excessively positive charged hydrogen atom covalently bonded with a donor. From the above description, a picture of a electrostatic interaction between dissimilar electric charges emerges, and in fact it is so. However, a hydrogen bond donor, allowing a hydrogen atom to present a positive charge, requires a shift of this charge within these two atoms, which introduces a covalent bond component to the problem of hydrogen bonding. Indeed, consideration of the mechanism of hydrogen bonding leads to the conclusion that it is a mixture of electrostatic and covalent interactions. The difficulties related to the proper interpretation of hydrogen bonding are well described in [13], and the International Union of Pure and Applied Chemistry (IUPAC) recommendation regarding the definition of this bond is available in the publication [14]. It is also worth knowing that the currently available sophisticated research tools enable imaging not only of covalent bonding orbitals, but also the location and presence of hydrogen bonds [15], while theoretical considerations strongly suggest the quantum effect of hydrogen confinement in the hydrogen bond, clearly visible experimentally in the difference in peak positions in the vibrational spectrum of the X–H group (for example, N–H or O–H) in the situations of non-confinement of H, i.e. without hydrogen bonds, and of quantum confinement of H, when the hydrogen bond exists [16, 17].

The universality and importance of the occurrence of hydrogen bonds is undeniable. It may be useful to state that the highest density of hydrogen bonds occurs in water, and these bonds allow water to appear in a liquid form under normal Earth conditions, and are responsible for most of the amazing properties of water [18]. Hydrogen bonds, although usually much weaker than covalent bonds, due to their commonness are responsible for the stability of three-dimensional molecular forms of nucleic acids [16, 19] and proteins [20, 21], molecular recognition

[22, 23], and also for the presence of matter in a condensed form, not a gas, as exemplified by the presence of hydrogen bonds in minerals with a layered structure [24]. The large number of relatively weak hydrogen bonds guarantees the stability of molecular structures, but also the ease of inducing and preserving changes in these structures by local breaking of existing structures and creating new hydrogen bonds (for example, denaturation and renaturation of DNA and proteins).

The remaining bonds, participating mainly in the intermolecular interactions, are electrostatic in nature and come from the following interactions, listed according to decreasing strength of the interactions. The first are ion–ion interactions, different from ionic bonds due to the lack of localization of electric charges on individual interacting atoms. In these cases, the charge is delocalized and belongs to several atoms of the molecule. The next group of bonds comes from an interaction of a permanent dipole–permanent dipole type and concerns molecules with an excess of positive charge in one place and negative charge in another. However, this does not apply to the hydrogen bonds discussed above (for example, water molecules are permanent electric dipoles, but they interact with each other to form an H-bond, which is stronger). Next, the interaction between a permanent dipole and a induced dipole belongs to the group of van der Waals interactions. This interaction results from the induction by a molecule that is a permanent dipole of electric charge shifts in a molecule that is electrically neutral and does not have a permanent dipole moment. This leads to interactions similar to the permanent dipole–permanent dipole type, but the interactions are much weaker. There are also dispersive interactions (London dispersive forces) resulting from the interactions between the induced dipoles. Dispersion interactions always occur between molecules, regardless of whether they have permanent dipole moments or not. This interaction results from the momentary asymmetric distribution of the electric charge of the electrons around the atomic nucleus, which by itself induces a momentary dipole moment of such an atom, which in turn can induce dipole moments in adjacent atoms. Although this interaction is one of the weakest, due to the commonness of its occurrence, its share in the resultant intermolecular interaction may exceed the share of dipole–dipole interactions. One should also remember the repulsion between atoms and molecules, not classified into the category of interactions, resulting from the Pauli principle of exclusion ordering the world on a submolecular and atomic scale.

Aside from the above considerations regarding electrostatic interactions, it should be noted that whenever we deal with a charge shift in an atom or molecule, one should be aware that it is a quantum mechanical process subject to all the rules of quantum physics, in particular the rules resulting from the Pauli principle of exclusion.

Tunneling of the potential barrier by an electron or an α particle is certainly not an isolated phenomenon that allows the reaction to take place. The hydrogen atom, or its ionized form, the proton, due to its low mass (many times, about two thousand times greater than the mass of the electron, but smaller than the mass of the α particle), is particularly predestined to tunnel the potential barriers represented by the activation energy. There are known reactions involving the transfer of hydrogen from one molecule to another [25, 26], as well as within the same molecule [27, 28]. Thus, the possibility of tunneling the potential barrier by hydrogen atoms or protons is naturally linked to the enzymatic activity of numerous enzymes.

Because the hydrogen atom can exist in three isotopic forms (protonium, deuterium, and tritium) significantly different in terms of the mass of atomic nuclei (protonium—proton; deuterium—proton and neutron; tritium—proton and two neutrons) with the same electron configuration in the atom, guaranteeing the same chemical reactivity, it became possible to experimentally verify the role of hydrogen tunneling in the enzymatic reactions taking place. The method of kinetic isotope effects (KIEs) resulting from the different probability of tunneling the potential barrier by individual hydrogen isotopes was used. In 1989 a report was published indicating significant deviations from the semiclassical predictions of kinetics of the oxidation of benzyl alcohol to benzyl aldehyde catalyzed by yeast alcohol dehydrogenase, but these deviations corresponded very well to the predictions resulting from taking into account the significant contribution of the tunneling effect [29]. A detailed analysis of the hydrogen tunneling mechanism in enzymatic reactions can be found in the papers [30, 31], and an interesting discussion on the possible evolution of enzymes towards the use of quantum mechanics is presented in the paper titled 'By chance, or by design?' [32]. These publications were followed by numerous studies confirming the presence of the hydrogen tunneling effect in enzymatic reactions and in-depth analyses of this process [33–36].

It is worth paying attention to one of the most important topics in molecular biology, which is the genetic stability of DNA, and one of the threats to this stability is the transition of protons along hydrogen bonds stabilizing the double stranded DNA structure. Such transitions of protons could lead to tautomerization of bases and as a result enable the formation of point mutations. The theoretical analysis of the hydrogen bonds between guanine–cytosine nucleotides (G–C) is presented in the recent work [37]. This analysis includes the exact model of the base pairs structure, the quantum dynamics of the proton in the hydrogen bond, and the influence of quantum decoherence by the environment in the cell. It has been shown that the speed of proton transfer using the tunneling process is several orders of magnitude greater than the expected classic transition over the energy barrier, and the estimated probability of such a process suggests that such proton tunneling can play a much greater role in DNA mutation than suggested before.

6.7 Molecular tunneling and molecular interactions

It might seem that the transfer of larger objects in the form of groups of atoms goes beyond the quantum mechanical tunneling of the potential barrier represented by the activation energy of the reaction. In fact, as the size and mass of objects increase, their quantum mechanical properties decrease sharply, but this does not mean that their movement within a molecule or between molecules is not controlled by quantum effects. It has been shown that the displacement of a large size of atomic moiety inside the enzyme, or the sequential transfer of one enzyme product as a substrate for another enzyme, is supported and controlled by the formation of intramolecular or intermolecular tunnels [38–41]. Such molecular tunnels facilitate the reaction and show high substrate specificity. Although the mechanism of the formation of molecular tunnels is not fully understood, their functioning seems to be different than in the case of tunneling the potential barrier by an electron or hydrogen atom, however, the formation of such structures and their active functioning seem to be related to quantum mechanisms involved in hydrogen bonding and much weaker but more common electrostatic dispersion interactions.

It is worth noting that despite the universality of tunneling effects, they cannot be fully responsible for the processes of formation and breaking of bonds between atoms and for enzymatic activity. Certainly, only meeting, even at a certain distance, of whole molecules or their distinguished atomic groups, results in significant changes in the electric charge density and the possibility of rearrangement of bonds between atoms. For example, the interaction of a signal molecule (hormone) with a receptor, or an antigen with an antibody, results in conformational changes of this receptor molecule and the initiation of the information transduction process in the chain of molecular interactions [42–44].

6.8 Interactions with nanomaterials

Equally significant conformational changes of proteins were observed in the case of their interaction with nanoparticles [45, 46], but it is worth noting that the sizes of nanoparticles or nanostructures may be comparable to the sizes of proteins. This is due to the fact that in the nanoparticle–protein interaction one should take into account their size ratio, which strongly influences the result of such interaction. If the nanoparticle is much larger than the protein molecule, then it can be assumed, due to the significant difference of the curvature radii of both objects, that the protein interacts with a substantially flat surface by multiple points contact. However, if the radius of curvature of the nanoparticle decreases at the constant size of the protein molecule, then the multi-point interaction with the curved surface introduces additional stresses in the protein structure, significantly increasing the observed conformational changes of this protein. The situation is even different when the protein molecule is much larger than the nanoparticle. Then the multi-point interaction is significantly reduced and conformational changes in the protein molecule may be of local importance. It has been assumed that the influence of the interaction of proteins with nanoparticles on the properties of nanoparticles resulting from the formation of the so-called protein corona on the nanoparticle surface. However, as it results from the above, the corona effect depends strongly on the ratio of the sizes of nanoparticles and protein molecules, and the contact of these structures also results in changes in the conformation and properties of protein molecules.

As can be seen from the above considerations, the interaction of the artificial nanoparticle surface, or the nano-structured macroscopic surface, induces conformational changes in the exposed protein molecule. These may be transient changes that disappear upon termination of contact, but may be responsible for activating the signal transduction pathway, such as with signal molecules, leading to a metabolic response of the cell. This response can be highly specific, which can be used to recognize the surface by cells as a source of specific stress activating the appropriate metabolic pathways in the cell.

However, the induced changes may be permanent, as is the case with prion disease that develops as a result of induction of conformational changes in the prion protein by contact with the pathologically conformationally altered prion protein [47].

On the margin of these considerations, it is also worth mentioning that it cannot be ruled out that the accidentally produced nanostructure causes similar conformational changes in receptor protein molecules as the signal molecule, which can be described as a biomimetic effect. Due to the high durability of such a biomimetic nanostructure, it can be useful for biomedical purposes. However, the case of a randomly produced nanostructure, potentially capable of inducing conformational changes leading to the emergence of a metabolic response in a pathological form leading to a dangerous disease, is also possible. It can also be a disease of unknown course, unknown etiology, and unknown treatment. Unfortunately, such an event cannot be ruled out at the present state of knowledge, therefore the biological assessment of innovative technology products should be very meticulous.

References

[1] Wong B 2019 On quantum entanglement *Int. J. Autom. Control Syst.* **5** 1–7

[2] Macdonald A 2012 Spooky action at a distance: the puzzle of entanglement in quantum theory http://faculty.luther.edu/~macdonal/Spooky.pdf?msclkid=1173459da9d511ec9cbdd582eb909c0a

[3] Liu L 2007 The quantum mechanics of alpha decay *MIT* http://web.mit.edu/lululiu/Public/pixx/not-pixx/alpha.pdf

[4] Trixler F 2013 Quantum tunnelling to the origin and evolution of life *Curr. Org. Chem.* **17** 1758–70

[5] Lower S 2004 Electron-tunneling model of chemical bonding *Chem1 Virtual Textbook* http://chem1.com/acad/webtut/bonding/TunnelBond.html

[6] Nordholm S 1988 Delocalization—the key concept of covalent bonding *J. Chem. Educ.* **65** 581–4

[7] Marcus R A 1956 On the theory of oxidation–reduction reactions involving electron transfer *J. Chem. Phys.* **24** 966–78

[8] Gray H B and Winkler J R 2005 Long-range electron transfer *Proc. Natl Acad. Sci.* **102** 3534–9

[9] Winkler J R and Gray H B 2014 Long-range electron tunneling *J. Am. Chem. Soc.* **136** 2930–9

[10] Fereiro J A, Yu X, Pecht I, Sheves M, Cuevas J C and Cahen D 2018 Tunneling explains efficient electron transport via protein junctions *Proc. Natl Acad. Sci.* **115** E4577–83

[11] Schreiner P R 2017 Tunneling control of chemical reactions: the third reactivity paradigm *J. Am. Chem. Soc.* **139** 15276–83

[12] Lagunas A, Guerra-Castellano A, Nin-Hill A, Díaz-Moreno I, De la Rosa M A, Samitier J, Rovira C and Gorostiza P 2018 Long distance electron transfer through the aqueous solution between redox partner proteins *Nat. Commun.* **9** 5157

[13] Desiraju G R 2011 A bond by any other name *Angew. Chem. Int. Ed.* **50** 52–9

[14] Arunan E *et al* 2011 Definition of the hydrogen bond (IUPAC recommendations 2011) *Pure Appl. Chem.* **83** 1637–41

[15] Zhang J, Chen P, Yuan B, Ji W, Cheng Z and Qiu X 2013 Real-space identification of intermolecular bonding with atomic force microscopy *Sci. Rep.* **342** 611–4

[16] dos Santos C S, Filho E D and Ricotta R M 2015 Quantum confinement in hydrogen bond of DNA and RNA *J. Phys. Conf. Ser.* **597** 012033

[17] dos Santos C S, Filho E D and Ricotta R M 2015 Quantum confinement in hydrogen bond *Int. J. Quantum Chem.* **115** 765–70

[18] Rye C, Wise R, Jurukovski V, DeSaix J, Choi J and Avissar Y 2013 Water *Biology* (Houston, TX: OpenStax) ch 2.2 pp 47–53 https://openstax.org/books/biology/pages/2-2-water

[19] Guerra C F, Bickelhaupt F M, Snijders J G and Baerend E J 2000 Hydrogen bonding in DNA base pairs: reconciliation of theory and experiment *J. Am. Chem. Soc.* **122** 4117–28

[20] Pace C N *et al* 2014 Contribution of hydrogen bonds to protein stability *Protein Sci.* **23** 652–61

[21] Bianco V, Iskrov S and Franzese G 2010 Effect of hydrogen bonds on protein stability, arXiv:1011.3785

[22] Jeffrey G A 1996 Hydrogen bonds and molecular recognition *Food Chem.* **56** 241–6

[23] Martinez-Felipe A, Brebner F, Zaton D, Concellon A, Ahmadi S, Piñol M and Oriol L 2018 Molecular recognition via hydrogen bonding in supramolecular complexes: a fourier transform infrared spectroscopy study *Molecules.* **23** 2278

[24] Conroy M, Soltis J A, Wittman R S, Smith F N, Chatterjee S, Zhang X, Ilton E S and Buck E C 2017 Importance of interlayer H bonding structure to the stability of layered minerals *Sci. Rep.* **7** 13274

[25] Cronholm T 1985 Hydrogen transfer between ethanol molecules during oxidoreduction *in vivo Biochem. J.* **229** 315–22

[26] Kobayashi N, Minami T, Tani A, Nakagoshi M, Sugahara T, Takeya K and Ohgaki K 2012 Intermolecular hydrogen transfer in isobutane hydrate *Energies* **5** 1705–12

[27] Piatkowski L, Schanbacher C, Wackenhut F, Jamrozik A, Meixner A J and Waluk J 2018 Nature of large temporal fluctuations of hydrogen transfer rates in single molecules *J. Phys. Chem. Lett.* **9** 1211–5

[28] Dabkowska I, Rak J, Gutowski M, Nilles J M, Stokes S T, Radisicd D and Bowen K H Jr 2004 Barrier-free proton transfer in anionic complex of thymine with glycine *Phys. Chem. Chem. Phys.* **6** 4351–7

[29] Cha Y, Murray C J and Klinman J P 1989 Hydrogen tunneling in enzyme reactions *Science* **243** 1325–30

[30] Antoniou D and Schwartz S D 1997 Large kinetic isotope effects in enzymatic proton transfer and the role of substrate oscillations *Proc. Natl Acad. Sci. USA* **94** 12360–5

[31] Sen A and Kohen A 2009 Quantum effects in enzyme kinetics *Quantum Tunnelling in Enzyme-Catalyzed Reactions* ed N Scrutton and R Allemann (London: RSC Publishing) ch 7 pp 161–78

[32] Ball P 2004 By chance, or by design? *Nature* **431** 396–7

[33] Nagel Z D and Klinman J P 2006 Tunneling and dynamics in enzymatic hydride transfer *Chem. Rev.* **106** 3095–118

[34] Pusuluk O, Farrow T, Deliduman C, Burnett K and Vedral V 2018 Proton tunnelling in hydrogen bonds and its implications in an induced-fit model of enzyme catalysis *Proc. R. Soc.* A **474** 0037

[35] Moser C C, Anderson J L R and Dutton. P L 2010 Guidelines for tunneling in enzymes *Biochim. Biophys. Acta* **1797** 1573–86

[36] Klinman J P and Kohen A 2013 Hydrogen tunneling links protein dynamics to enzyme catalysis *Annu. Rev. Biochem.* **82** 471–96

[37] Slocombe L, Sacchi M and Al-Khalili J 2022 An open quantum systems approach to proton tunnelling in DNA *Commun. Phys.* **5** 109

[38] Hyde C C, Ahmed S A, Padlan E A, Miles E W and Davies D R 1988 Three-dimensional structure of the tryptophan synthase $\alpha 2\beta 2$ multienzyme complex from *Salmonella typhimurium J. Biol. Chem.* **263** 17857–87

[39] Raushel F M, Thoden J B and Holden H M 2003 Enzymes with molecular tunnels *Acc. Chem. Res.* **36** 539–48

[40] Banerjee R and Lipscomb J D 2021 Small-molecule tunnels in metalloenzymes viewed as extensions of the active site *Acc. Chem. Res.* **54** 2185–95

[41] Singh S and Anand R 2021 Tunnel architectures in enzyme systems that transport gaseous substrates *ACS Omega* **6** 33274–83

[42] Schmidt A, MacColl R, Lindau-Shepard B, Buckler D R and Dias J A 2001 Hormone-induced conformational change of the purified soluble hormone binding domain of follitropin receptor complexed with single chain follitropin *J. Biol. Chem.* **276** 23373–81

[43] Ward C W, Menting J G and Lawrence M C 2013 The insulin receptor changes conformation in unforeseen ways on ligand binding: sharpening the picture of insulin receptor activation *Bioessays* **35** 945–54

[44] Al Qaraghuli M M, Kubiak-Ossowska K, Ferro V A and Mulheran. P A 2020 Antibody–protein binding and conformational changes: identifying allosteric signalling pathways to engineer a better effector response *Sci. Rep.* **10** 13696

[45] Simón-Vázquez R, Lozano-Fernández T, Peleteiro-Olmedo M and González-Fernández Á 2014 Conformational changes in human plasma proteins induced by metal oxide nanoparticles *Colloids Surf.* B **113** 198–206

[46] Park S J 2020 Protein–nanoparticle interaction: corona formation and conformational changes in proteins on nanoparticles *Int. J. Nanomed.* **15** 5783–802

[47] McKintosh E, Tabrizi S J and Collinge J 2003 Prion diseases *J. NeuroVirol.* **9** 183–93

Chapter 7

Summary

This book presents and discusses our own analysis of the history of human thought and science against the background of the development of the world, as well as the presentation of the latest achievements of science and technology based on this analysis in the field of materials science and life sciences and their mutual interpenetration and stimulation of development.

One of the central concepts discussed on the pages of this books is *materiomics*, understood as a field of co-operation, exchange of experiences, as well as setting common goals and indicating the ways of their effective achievement to meet economic and social needs. Materiomics uses the experience of materials science and the life sciences and, by collecting data in the field of the production, use, and processing of new materials using physicochemical and biological methods as well as IT techniques, enables comprehensive use of these data (the *materiom*) to develop innovative material solutions, including the possibility of personalizing products.

One should be aware that the matter around us, both inanimate and animate, consist of the same components in the form of atoms and elementary particles, subject to the same laws of physics and chemistry, including the creation of complex molecular structures and their mutual interactions. However, unlike inanimate matter, animate matter has an amazing property consisting in the ability to organize itself and reproducing itself. The secret of this property is contained in the form of genetic information accumulated in the super biomaterial, which is genetic material in the form of the DNA molecule. Despite the enormous progress of molecular biology methods, the ability to fully read and use this information, available at the molecular level in the cell, remains not fully available to us and is full of secrets. It seems that with the progress in knowledge of these properties of animate matter, a technological breakthrough in the area of the design, production, and use of products having the attributes of inanimate matter will also be made.

Turning to nanotechnology seems to have unlimited prospects for technological development. It should be remembered, however, that these extraordinary properties

doi:10.1088/978-0-7503-2656-8ch7

of nanomaterials (nanoparticles, nanostructures) result from the universality of quantum effects that play a dominant role on a nanometric scale. With the increase in the size of the object, its quantum features disappear and the object becomes a macroscopic object available to our senses and imagination of matter. It is not understandable why nanometric objects created as a result of natural processes and engineering procedures are still distinguished from nanometric objects crucial for the functioning of living organisms, i.e. protein particles, nucleic acids, fatty acids or polysaccharides. After all, these molecules consist of the same atoms and are subject to the same laws of physics and chemistry, including the universality of quantum effects, as naturally created or as a result of engineering activities. In fact, we can produce these particles by engineering methods during their *in vitro* extracellular synthesis, not mentioning the achievements of genetic engineering allowing for the production of recombinant proteins that already form a huge percentage of medicinal substances. After all, this is also nanotechnology, in the latter case using biotechnological methods.

What's more, biological objects can 'see', 'feel', and recognize nanomaterials/nanostructures when encountering them and as a result start specific metabolic pathways, literally in the same way as this happens in the case of contact with signal molecules and other substances such as pheromones, fragrance and flavor molecules, medicines, and poisons. This makes cells, as well as entire multicellular organisms, extremely sensitive and selective sensors of the presence of nanomaterials/nanostructures. A deeper knowledge of the nanomaterial–metabolic response relationship of living organisms will enable us to precisely identify nanomaterials/nanostructures and may allow for effective grouping of these materials, as atoms are managed in the table of periodicity. We devoted a significant part of this book to discussing the issues of molecular recognition and, as a result, we propose using existing molecular biology tools (transcriptomics and proteomics) for this purpose. We presented preliminary analyzes of similarities and differences in transcriptome profiles using the correlation analysis of these profiles.

As a complementary approach to the identification of nanomaterials/nanostructures, we also proposed the use of existing tools of experimental physics in the form of STM microscopy for imaging and identification of orbitals as well as photoelectron spectrometry (PES) to assess the eigen-energy of electrons. We have indicated potential problems to solve when using STM microscopy and how to use photoelectron energy spectra for comparative analysis.

The approach we proposed to identify nanomaterials/nanostructures can be described as non-standard analysis of results obtained by standard research methods of quantum physics and molecular biology. In our approach, we are not interested in in-depth analyzes of the structure of the nanoparticles and electron states of the atoms in it, or in-depth analysis of processes occurring in the cell as a result of contact with nanomaterial/nanostructure. We suggest the use of similarities and differences in PES energy spectra as well as similarities and differences in transcriptome and proteome profiles for the identification of nanomaterials/nanostructures. The right selection of methods for analyzing these similarities and differences remains the difficulty of this approach. At the same time, we do not limit the possibility of using the obtained experimental results for in-depth structural and metabolic analyzes.

We would like to pay separate attention to the need to use a quantum mechanical description of the interactions of atoms and molecules for the purposes of life sciences. Publications devoted to these issues are increasingly appearing, but the concept of quantum biology is still associated mainly with biological processes caused by photons carrying energy in a quantized form. It is rare to indicate the identity of animate and inanimate matter from the point of view of the atomic structure of matter and the quantum mechanical laws in force on this scale. Perhaps adopting this point of view will bring us closer to the secret of life, as long as this secret is available to us at the current level of knowledge.

We thank you for the time you have spent reading and analyzing the content of this book. We remain in the hope that the point of view we presented on the history and future of science interested the reader and stimulated the independent analysis of available data and an interdisciplinary approach to conducted or planned research work.

Lodz/Dobra-Nowiny, 11 May 2022

www.ingramcontent.com/pod-product-compliance
Ingram Content Group UK Ltd.
Pitfield, Milton Keynes, MK11 3LW, UK
UKHW051849100926
8556IPUK00005B/533

9 780750 326544